AGRICULTURE ISSUES AND POLICIES

FOREST SERVICE TRAILS

MAINTENANCE AND SUSTAINABILITY ISSUES

AGRICULTURE ISSUES AND POLICIES

Additional books in this series can be found on Nova's website under the Series tab.

Additional E-books in this series can be found on Nova's website under the E-book tab.

ENVIRONMENTAL RESEARCH ADVANCES

Additional books in this series can be found on Nova's website under the Series tab.

Additional E-books in this series can be found on Nova's website under the E-book tab.

AGRICULTURE ISSUES AND POLICIES

FOREST SERVICE TRAILS

MAINTENANCE AND SUSTAINABILITY ISSUES

ADAM D. FAYDEN
EDITOR

NOVINKA

New York

For permission to use material from this book please contact us:
Telephone 631-231-7269; Fax 631-231-8175
Web Site: http://www.novapublishers.com

Library of Congress Cataloging-in-Publication Data

ISBN: 978-1-62808-834-2

Published by Nova Science Publishers, Inc. † New York

CONTENTS

Preface vii

Chapter 1 Forest Service Trails: Long- and Short-Term Improvements Could Reduce Maintenance Backlog and Enhance System Sustainability 1
United States Government Accountability Office

Chapter 2 American Recovery and Reinvestment Act – Forest Service Capital Improvement and Maintenance Projects – Trail Maintenance and Decommissioning 57
United States Department of Agriculture

Index 91

PREFACE

The Forest Service manages more than 158,000 miles of recreational trails offering hikers, horseback riders, cyclists, off-highway-vehicle drivers, and others access to national forests. To remain safe and usable, these trails need regular maintenance, such as removal of downed trees or bridge repairs. GAO was asked to review the agency's trail maintenance activities.

Chapter 1 - This report examines (1) the extent to which the Forest Service is meeting trail maintenance needs, and effects associated with any maintenance not done; (2) resources, including funding and labor, that the agency employs to maintain its trails; (3) factors, if any, complicating agency efforts to maintain its trails; and (4) options, if any, that could improve the agency's trail maintenance efforts. GAO reviewed laws and agency documents; analyzed Forest Service budget data for fiscal years 2006-2012 and trails data for fiscal years 2008-2012; and interviewed agency officials and representatives of 16 stakeholder groups selected to represent trail users, conservation, and industry. Their views are not generalizable.

Chapter 2 – To create jobs and promote economic growth, Congress passed the American Recovery and Reinvestment Act (Recovery Act) on February 17, 2009. The Recovery Act provided Forest Service (FS) $1.15 billion—including $650 million for its Capital Improvement and Maintenance (CIM) fund, and $500 million for its Wildland Fire Management (WFM) fund—to fund projects that directly accomplish FS' mission of sustaining the nation's forests and grasslands. FS allocated $99 million in Recovery Act funds to 90 trail maintenance and decommissioning projects (trail projects). To administer these projects, FS awarded the funds primarily through contracts and partnership agreements. For partnership agreements, FS tended to choose local cooperators (public, private, and non-profit organizations), many of which had an existing partnership with FS.

Congress, in enacting the Recovery Act, emphasized the need for accountability and transparency in the expenditure of funds. Furthermore, the Office of Management and Budget (OMB) issued initial and additional guidance that required Federal agencies to establish rigorous internal controls, oversight mechanisms, and other approaches to meet the accountability objectives of the Recovery Act. The Recovery Act mandated that the U.S. Department of Agriculture's Office of Inspector General (OIG) oversee agency activities to ensure Recovery Act funds were spent in a manner that minimized the risk of improper use. We, therefore, conducted an audit to evaluate FS' administration of the Trail Maintenance and Decommissioning program funds.

In: Forest Service Trails
Editor: Adam D. Fayden
ISBN: 978-1-62808-834-2

Chapter 1

FOREST SERVICE TRAILS: LONG- AND SHORT-TERM IMPROVEMENTS COULD REDUCE MAINTENANCE BACKLOG AND ENHANCE SYSTEM SUSTAINABILITY*

United States Government Accountability Office

WHY GAO DID THIS STUDY

The Forest Service manages more than 158,000 miles of recreational trails offering hikers, horseback riders, cyclists, off-highway-vehicle drivers, and others access to national forests. To remain safe and usable, these trails need regular maintenance, such as removal of downed trees or bridge repairs. GAO was asked to review the agency's trail maintenance activities. This report examines (1) the extent to which the Forest Service is meeting trail maintenance needs, and effects associated with any maintenance not done; (2) resources, including funding and labor, that the agency employs to maintain its trails; (3) factors, if any, complicating agency efforts to maintain its trails; and (4) options, if any, that could improve the agency's trail maintenance efforts. GAO reviewed laws and agency documents; analyzed Forest Service budget data for fiscal years 2006-2012 and trails data for fiscal years 2008-2012; and

* This is an edited, reformatted and augmented version of United States Government Accountability Office, Publication No. GAO-13-618, dated June 2013.

interviewed agency officials and representatives of 16 stakeholder groups selected to represent trail users, conservation, and industry. Their views are not generalizable.

What GAO Recommends

GAO recommends, among other actions, that the Forest Service (1) analyze trails program needs and available resources and develop options for narrowing the gap between them and take steps to assess and improve the sustainability of its trails and (2) take steps to enhance training on collaborating with and managing volunteers who help maintain trails. In commenting on a draft of this report, the Forest Service generally agreed with GAO's findings and recommendations.

What GAO Found

The Forest Service has more miles of trail than it has been able to maintain, resulting in a persistent maintenance backlog with a range of negative effects. In fiscal year 2012, the agency reported that it accomplished at least some maintenance on about 37 percent of its 158,000 trail miles and that about one-quarter of its trail miles met the agency's standards. The Forest Service estimated the value of its trail maintenance backlog to be $314 million in fiscal year 2012, with an additional $210 million for annual maintenance, capital improvement, and operations. Trails not maintained to quality standards have a range of negative effects, such as inhibiting trail use and harming natural resources, and deferring maintenance can add to maintenance costs.

The Forest Service relies on a combination of internal and external resources to help maintain its trail system. Internal resources include about $80 million allocated annually for trail maintenance activities plus funding for other agency programs that involve trails. External resources include volunteer labor, which the Forest Service valued at $26 million in fiscal year 2012, and funding from federal programs, states, and other sources.

Collectively, agency officials and stakeholders GAO spoke with identified a number of factors complicating the Forest Service's trail maintenance efforts, including (1) factors associated with the origin and location of trails,

(2) some agency policies and procedures, and (3) factors associated with the management of volunteers and other external resources. For example, many trails were created for purposes other than recreation, such as access for timber harvesting or firefighting, and some were built on steep slopes, leaving unsustainable, erosion-prone trails that require continual maintenance. In addition, certain agency policies and procedures complicate trail maintenance efforts, such as the agency's lack of standardized training in trails field skills, which limits agency expertise. Further, while volunteers are important to the agency's trail maintenance efforts, managing volunteers can decrease the time officials can spend performing on-the-ground maintenance.

Agency officials and stakeholders GAO interviewed collectively identified numerous options to improve Forest Service trail maintenance, including (1) assessing the sustainability of the trail system, (2) improving agency policies and procedures, and (3) improving management of volunteers and other external resources. In a 2010 document titled *A Framework for Sustainable Recreation*, the Forest Service noted the importance of analyzing recreation program needs and available resources and assessing potential ways to narrow the gap between them, which the agency has not yet done for its trails. Many officials and stakeholders suggested that the agency systematically assess its trail system to identify ways to reduce the gap and improve trail system sustainability. They also identified other options for improving management of volunteers. For example, while the agency's goal in the *Forest Service Manual* is to use volunteers, the agency has not established collaboration with and management of volunteers who help maintain trails as clear expectations for trails staff responsible for working with volunteers, and training in this area is limited. Some agency officials and stakeholders stated that training on how to collaborate with and manage volunteers would enhance the agency's ability to capitalize on this resource.

ABBREVIATION

OHV	off-highway vehicle

United States Government Accountability Office
441 G St. N.W.
Washington, DC 20548
June 27, 2013

The Honorable Michael K. Simpson
Chairman

The Honorable James P. Moran
Ranking Member
Subcommittee on Interior, Environment, and Related Agencies
Committee on Appropriations
House of Representatives

The Honorable Cynthia M. Lummis
House of Representatives

Growing populations, expanding development, and the lure of the outdoors for people living near public lands have increased demand for recreational opportunities in national forests and grasslands, including demand for trails. Recreation enthusiasts contend that trails enhance quality of life, and, according to the Outdoor Industry Association, the use of trails contributes over $80 billion annually to the U.S. economy. The Forest Service, within the Department of Agriculture, manages more than 158,000 miles of recreational trails on its 155 national forests and 20 national grasslands—the largest network of trails in the country, according to the agency. Forest Service data show that national forests receive about 165 million visits each year, and many of these visits involve trails for a variety of recreational pursuits. The primary recreation activity on trails is recreation on foot, with over 40 percent of visitors to national forests hiking or walking the forests' trails. In addition, over the last 40 years, new forms of motorized recreation, such as snowmobiles and off-highway vehicles (OHV), as well as nonmotorized recreation, such as mountain bikes, have gained popularity. These new forms of recreation intersect with many traditionally popular activities, such as camping, hiking, hunting, and horseback riding.

Recreational trails require regular maintenance, including removal of trees and other debris and repairs to bridges. Without proper maintenance, trails can become dangerous or unusable, and deferring needed maintenance can ultimately add to costs. In 1989, we reported a nearly $200 million backlog of deferred maintenance on Forest Service trails, noting that the agency did not routinely gather data on maintenance and reconstruction needs or associated costs. Additionally, we found that the Forest Service's new computerized information system, to be operational in 1990, was not planned to gather such data.[1] Since we issued that report, the Forest Service has faced new trail maintenance challenges, including widespread wildland fires, extensive tree die-offs due to insect and disease outbreaks, and more recreational use of its trails, all of which can require more maintenance to keep trails safe and passable. In light of these and other factors affecting trail maintenance, along with constrained federal budgets, trail users and others have remained concerned about the Forest Service's ability to adequately maintain its trail system.

In this context, you asked us to review the Forest Service's trail maintenance activities. This report examines (1) the extent to which the Forest Service is meeting trail maintenance needs, and effects associated with any maintenance not done; (2) resources, including funding and labor, that the agency employs to maintain its trails; (3) factors, if any, complicating agency efforts to maintain its trails; and (4) options, if any, that could improve the agency's trail maintenance efforts.

To conduct this work, we reviewed relevant laws and agency documents, including agency handbooks and other guidance related to maintaining trails. We interviewed Forest Service officials in the agency's headquarters and received information from all nine of its regions about trail maintenance needs and effects associated with any deferred maintenance. We also interviewed officials from a nonprobability sample of 18 national forests located in five of the nine Forest Service regions. We held these interviews to learn about their trail maintenance programs; we also examined trails on which maintenance had been deferred and trails that were well maintained. We selected these forests to represent variation in geography, proximity to urban and rural areas, trail mileage, and type and intensity of trail use, although findings from this selection of forests are not generalizable to the entire population of national forests. We obtained data on the Forest Service's trail inventory for fiscal years 2008 to 2012 from the agency's Infrastructure database (known as Infra).

We assessed the reliability of these data by reviewing relevant documentation and interviewing agency officials knowledgeable about the data and found them to be sufficiently reliable for the purposes of this report. To evaluate the resources the Forest Service employs to maintain its trails, we reviewed allocation data in agency budget documents for fiscal years 2006 to 2012. We also collected and reviewed evidence from national, regional, forest, and ranger district officials about how funds are allocated for trail maintenance activities. In addition, we examined the agency's use of external resources in conducting trail maintenance, including laws, regulations, and agency guidance regarding the Forest Service's authority to use external resources. To obtain information on factors, if any, affecting trail maintenance efforts and options that could improve such efforts, we asked agency officials at all levels about both topics. Further, we convened a discussion group to gather the perspectives of Forest Service officials representing all nine regions on what options, if any, could improve trail maintenance efforts. We also selected a nonprobability sample of 16 organizations representing a variety of trail user, conservation, and industry perspectives and interviewed representatives of these organizations about their views on Forest Service trail conditions, factors influencing trail maintenance by the agency, and options related to maintenance efforts. The views of these stakeholders are not generalizable to all trail user, conservation, and industry organizations, but they provided various perspectives on the Forest Service's trail maintenance efforts. We also interviewed officials from three other federal land management agencies—the Department of the Interior's Bureau of Land Management, Fish and Wildlife Service, and National Park Service—to learn about their trail maintenance programs, as well as Interior's U.S. Geological Survey to learn about current research on trail design. (See app. I for further details on our objectives, scope, and methodology.)

We conducted this performance audit from June 2012 to June 2013 in accordance with generally accepted government auditing standards. Those standards require that we plan and perform the audit to obtain sufficient, appropriate evidence to provide a reasonable basis for our findings and conclusions based on our audit objectives. We believe that the evidence obtained provides a reasonable basis for our findings and conclusions based on our audit objectives.

BACKGROUND

The 193 million acres of public land managed by the Forest Service as national forests and grasslands are collectively known as the National Forest System. These lands are located in 44 states, Puerto Rico, and the Virgin Islands and make up about 9 percent of the United States' total land area (see figure 1).

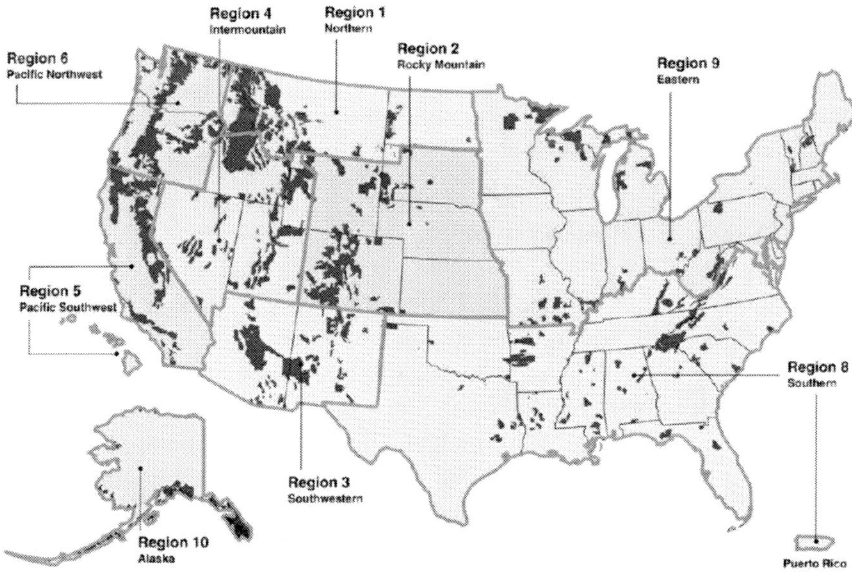

Source: Forest Service.
Note: The Forest Service does not have a region 7.

Figure 1. Forest Service Lands and Regions.

Forest Service's Organization

Stewardship of the National Forest System is carried out through nine regions that oversee 155 national forests; the forests, in turn, oversee more than 600 ranger districts. Each region encompasses a broad geographic area and is headed by a regional forester, who reports directly to the Chief of the Forest Service and provides leadership for, and coordinates the activities of, the various forests within the region. Each forest is headed by a supervisor, who allocates the budget and coordinates activities among the various ranger districts within the forest. Ranger districts, in turn, are headed by a district ranger, who conducts or oversees on-the-ground activities such as construction and maintenance of trails; operation of campgrounds; management of wildlife habitat; and the sale and harvest of forest products, including timber. Ranger districts vary in size from 50,000 acres to more than 1 million acres. Collectively, these field units are overseen by the Chief of the Forest Service, who operates out of the Forest Service's national headquarters in Washington, D.C. The Chief and other headquarters officials provide broad policy and direction for the agency, monitor the agency's activities, and inform Congress about agency accomplishments. In fiscal year 2012, the Forest Service had nearly 34,000 full-time-equivalent employees, about 97 percent of whom were in the field, and an enacted budget of about $5.6 billion.

Forest Service's Trail System

At the close of fiscal year 2012, the Forest Service reported having about 158,000 miles of trail used for both recreation and management.[2] (See table 4 in app. II for information on the Forest Service's trail mileage, usage, and visitors.) Under the National Forest Management Act of 1976,[3] the Forest Service manages its lands for multiple uses—such as timber harvesting, watershed and wilderness protection, protection of fish and wildlife habitat, forage for livestock, and recreation—and the agency's trails provide access both for agency officials managing lands and for people visiting those lands. Located throughout Forest Service lands, these trails include many that existed before national forests were established and are managed under various land management authorities. For example, the Forest Service manages about 32,000 miles of trail in designated wilderness areas, which, under the Wilderness Act of 1964, are to be administered so as to leave them unimpaired for future use and enjoyment and to protect and preserve their wilderness character, among other goals.[4] Trails in wilderness areas are thus usually less developed and more rugged than nonwilderness trails. The Forest Service's trail system also includes parts of national scenic and historic trails established under the National Trails System Act of 1968.[5] These long, national scenic trails—such as the Appalachian and Pacific Crest Trails—are to "provide for maximum outdoor recreation potential and for the conservation and enjoyment of . . . the area through which such trails may pass."[6] National historic trails, such as the Oregon Trail, closely follow a historic travel route of national significance.

The Forest Service's trails program aims to ensure recreation opportunities, public safety, and backcountry access through operation, maintenance, rehabilitation, and improvement of forest trails. Forest Service trails are categorized by trail type, trail class, and the managed use of each trail. Trail type reflects predominant trail surface and general mode of travel for each trail. The three trail types are standard (or "terra") trails, which have a surface consisting predominantly of earth; snow trails, which have a surface consisting predominantly of snow or ice; and water trails, which have a surface consisting predominantly of water (but may include portage routes over land). The majority of Forest Service trails are terra trails, and in some cases, a trail may be classified as a terra trail in the summer and a snow trail in the winter. All Forest Service trails must also be categorized by trail class, which are general categories reflecting the prescribed scale of development for each trail. Specifically, class 1 trails are minimally developed, such as those with natural fords instead of bridges in wilderness areas, and are designed to provide a challenging recreation opportunity, usually in a natural and unmodified setting. Conversely, class 5 trails, such as those found at visitor centers or high-use recreation sites, are fully developed, have gentle grades, and are often paved. About half of National Forest System trails are class 3 trails, which may have some minor obstacles, such as rocks, and generally pose a moderate level of challenge to users. (For more information on miles of trails by trail class, see table 5 in app. II.) All Forest Service trails must have at least one managed use, which reflects the mode(s) of travel appropriate on a trail, given its design and management. For example, a trail may be designed and actively managed for hiker and equestrian use, although other uses, such as bicycling, might be allowed.

Information on a trail's type, class, use, and related design parameters is applied by land managers to set trail management objectives, which document each trail's intended purpose and how it is to be managed. Forest Service trails are to be maintained to the agency's national quality standards for trails, which describe conditions that trail users can expect to encounter and the level of trail quality the Forest Service plans to provide. For example, the standards state that trails and trailsides will be free of litter and human waste.

Trail Maintenance Reporting

Maintenance to keep trails in good condition may include, among other tasks, clearing encroaching vegetation and fallen trees, as well as repair; preventive maintenance; and replacement of trail signs, water drainage features, trail bridges, and other trail structures. For reporting purposes, the agency divides trail maintenance activities into three categories: (1) miles maintained, (2) miles meeting standard, and (3) miles improved. The Forest Service defines these categories as follows:

- *Miles maintained:* includes miles of trail on which at least one maintenance task was performed to quality standards during a given year, indicating that one or more—but not necessarily all—needed maintenance tasks were completed.

- *Miles meeting standard:* includes all trail miles that meet quality standards and have been maintained in accordance with a specific maintenance cycle associated with each trail's management objective. Maintenance cycles vary by trail; some trails, for example, may be on annual maintenance cycles, and others may be on 3- or 5-year cycles. Thus, a trail can meet the Forest Service's standards even if it was not maintained in a given year.
- *Miles improved:* includes all trail miles where any improvements were made during a given year through activities such as widening the trail and adding or improving trail bridges or trail components, such as barriers, trail surfacing, kiosks, and wildlife viewing platforms.

The Forest Service sets performance targets for miles maintained and miles improved, and collates accomplishment data from local units, including national forests or ranger districts, and reports data for each category in the agency's annual budget justification to Congress.

Forest Service's Authority to Use External Resources

In addition to using its own appropriations and staffing, the Forest Service is authorized to use volunteer labor and nonfederal funds in carrying out trail maintenance activities. Specifically, the Volunteers in the National Forests Act of 1972 authorizes the Forest Service to recruit, train, and accept the services of volunteers for a variety of activities related to national forests, including trail maintenance.[7] The agency may provide these volunteers transportation, uniforms, lodging, and subsistence support. The National Trails System Act also authorizes federal agencies, including the Forest Service, to encourage volunteer and volunteer organization involvement in the planning, development, maintenance, and management of trails, where appropriate. Under this act, volunteer work may include operating programs to organize and supervise volunteer trail-building efforts; conducting trail-related research projects; or educating and training volunteers on methods of trail planning, construction, and maintenance. Agencies are also authorized to provide volunteers with equipment, tools, and technical assistance.

In working with volunteers, the Forest Service generally uses two different types of agreements to outline expectations and address liability:

volunteer agreements or challenge cost-share agreements. [8] Volunteer agreements are signed by individuals or groups who are interested primarily in volunteering on a particular project or activity—for example, a local organization sponsoring a volunteer day for its members; such volunteers constitute the majority of volunteers to the Forest Service. Under a volunteer agreement, the Forest Service assumes liability for these individuals while they are volunteering, which authorizes the volunteers to collect workers' compensation if they are injured on the job.[9] Other volunteers are covered under challenge cost-share agreements, which outline the relationship between the Forest Service and a partner organization, identifying an exchange of funds or services between the agency and the partner group.[10] In this type of agreement, the partner organization certifies that it has liability insurance covering its volunteers. Generally, this type of agreement is used with certain organizations having long-standing relationships with the agency, such as youth and conservation corps.[11]

In addition to having the authority to accept volunteer labor, the Forest Service has authority to accept and use nonfederal funds to support trail maintenance. The Cooperative Funds Act authorizes the Forest Service to accept money received as contributions toward cooperative work in forest investigations or protection, management, and improvement of the National Forest System.[12] Under the act, the Forest Service may also apply for and receive grants under certain circumstances.

Travel Management Planning Process

The Forest Service has undertaken a large planning effort regarding the use of recreational motor vehicles in national forests and grasslands. Each national forest is to identify the minimum road system needed for safe and efficient travel and for administration, use, and protection of the National Forest System; roads that are no longer needed are to be decommissioned or considered for other uses, such as for trails. In addition, in 2005, the Forest Service promulgated a regulation known as the travel management rule, which, among other things, requires each national forest and grassland to identify and designate the roads, trails, and areas open to motor vehicles. In deciding whether to designate trails for motor vehicle use, the rule directs the Forest Service to consider, among other criteria, the need for and availability of resources to maintain and administer the trail if it were designated.

Forest Service Has Substantial Unmet Trail Maintenance Needs, Resulting in Multiple Negative Effects

The Forest Service has more miles of trail than it has been able to maintain, resulting in a long-standing deferred maintenance backlog. Trails not maintained to the Forest Service's standards may inhibit trail use and harm natural resources, and deferred maintenance can lead to increased maintenance costs in the future.

More Trails Exist Than Are Being Maintained, Leading to a Persistent Backlog in the Hundreds of Millions of Dollars

The Forest Service is unable to regularly maintain many of its 158,000 miles of trails. According to Forest Service data, over the last 5 years the agency performed at least some maintenance on an average of about one-third of its trail miles annually, with officials telling us that some trails had not received any maintenance in the last 10 years. For fiscal year 2012, the agency reported that it accomplished at least some maintenance on about 37 percent of its trail miles, or 59,274 miles of trail, exceeding its fiscal year 2012 target of 46,580 miles. Maintenance conducted ranged from minimal maintenance, such as pruning brush, to more extensive maintenance, such as repairing a bridge. In addition to maintenance, the agency improved about 1 percent of its trail miles each year over the last 5 years. Improvements could include, for example, adding platforms or upgrading trail surfaces. According to an agency official, the agency focuses more on conducting needed maintenance than on improving existing trails or constructing new ones. Over the past 5 years, from 17 to 41 percent of overall trail miles met Forest Service standards each year, with 26 percent (or about one-quarter) of trail miles meeting standards in fiscal year 2012. Figure 2 shows mileage totals for various measures relating to maintenance conducted and trail conditions over the past 5 fiscal years.

The lack of annual maintenance has led to a persistent deferred trail maintenance backlog, whose value in fiscal year 2012 was estimated by the Forest Service at $314 million. The Forest Service estimated an additional $210 million for that year in three other trail maintenance-related needs: annual maintenance, capital improvement, and operations. Together, these four estimates—deferred maintenance, annual maintenance, capital improvement, and operations—constitute the agency's annual estimate of its trail maintenance needs, which totaled about $524 million in fiscal year 2012 (see table 1).[13]

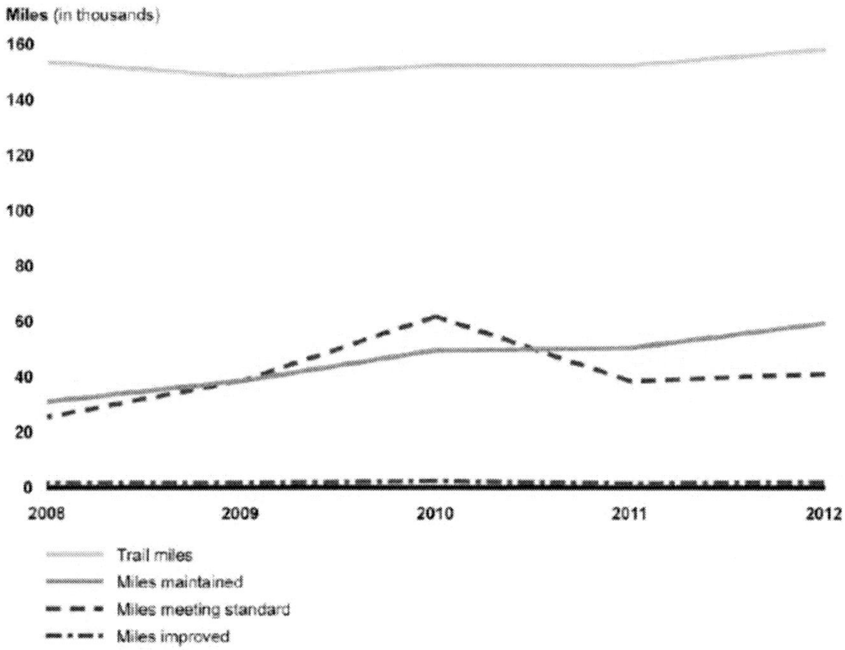

Source: GAO analysis of Forest Service data.

Figure 2. Forest Service Trail Miles, Miles Maintained, Miles Meeting Forest Service's Standards, and Miles Improved, Fiscal Years 2008-2012.

These estimates, however, may understate the scale of the agency's maintenance needs. Estimates are based on trail condition surveys conducted by local Forest Service staff on a random sample of approximately 1 percent of the agency's trail miles each year—the minimum number of trail miles that the agency has determined is required to generate a statistically valid estimate of its maintenance needs.[14] Some staff we interviewed, however, told us they do not always complete the surveys or ensure that they are providing accurate information for all trails included in the sample. They cited a number of difficulties associated with carrying out the surveys, including lack of available or trained personnel and a cumbersome and inefficient process that requires the surveyor to use a land-measuring wheel to measure the length of the trail and to carry a data dictionary while manually recording trail data.[15]

Forest Service headquarters officials told us they were taking steps to streamline the data collection process; these steps are discussed later in this report.

Table 1. The Forest Service's Estimated Trail Maintenance Needs for Fiscal Years 2006-2012

Year	Deferred maintenance[a]	Annual maintenance[b]	Capital improvement[c]	Operations[d]	Total
2006	$245,340,617	$53,056,992	$186,147,732	$24,590,348	$509,135,689
2007	224,165,294	52,620,623	128,443,693	16,574,566	421,804,176
2008	279,890,135	69,718,156	179,626,539	27,294,242	556,529,072
2009	293,954,942	69,661,704	165,123,504	22,088,815	550,828,965
2010	295,565,533	60,286,784	122,193,855	23,344,787	501,390,959
2011	295,744,022	66,573,536	114,663,581	21,090,602	498,071,741
2012	314,177,808	66,155,852	121,059,392	22,318,420	$523,711,472

Source: Forest Service.

[a]Deferred maintenance is maintenance that was not performed when it should have been, including repair, rehabilitation, or replacement of an asset—in this case, trails—to restore it to serviceable condition.

[b]Annual maintenance includes repair, preventive maintenance, or cyclic maintenance needed to maintain serviceability.

[c]Capital improvement refers to new construction, alteration, changing a trail's original function (e.g., changing from a hiking trail to an all-terrain-vehicle trail), or expanding or changing a trail's capacity.

[d]Operations refers to the Forest Service's estimate of annual operations costs for the trail maintenance program.

Unmaintained Trails Inhibit Trail Use, Harm Natural Resources, and Add to Agency Costs

Trails not maintained to the Forest Service's standards have a range of negative effects, including inhibiting trail use and posing potential safety hazards, harming natural resources, and adding to agency costs. Among the 18 national forests included in our review, officials at 15 forests cited various negative effects on visitors; officials from 10 forests specifically cited potential safety hazards as a consequence of deferred maintenance. For example, fallen logs across trails can impede hikers or block horseback, mountain bike, or OHV riders entirely (see figure 3). Officials from one forest noted that a safety hazard could arise from their inability to remove standing dead trees along a trail. Officials from another forest said that trail bridges needing replacement could be hazardous (see Figure 4), and officials at two other ranger districts cited concerns that users could get lost attempting to follow overgrown trails. Most forests we visited did not have trails that were closed because of deferred maintenance, but officials from a number of forests noted that they had some trails that were "functionally closed" because they were so overgrown or crowded with downed trees. Officials from several forests indicated that they had installed signs at trailheads warning of potential hazards. Outside the agency, nearly all the stakeholders we interviewed said they were concerned with the condition of the Forest Service's trail system and the agency's inability to maintain it adequately.

Source: Forest Service.

Figure 3. Trail Before and After Clearing Following a Windstorm, Nez Perce-Clearwater National Forests, Idaho, 2011.

Source: Forest Service.

Figure 4. Example of a Failing Bridge, Beaverhead-Deerlodge National Forest, Montana.

Unmaintained trails can also harm natural resources. For example, according to officials we interviewed at several forests, erosion resulting from unmaintained trails can create ecological damage. Trails with poor or unmaintained drainage features can deposit sediment into streams, degrading water quality and potentially affecting species, such as cutthroat trout. Officials at one forest stated that deferred maintenance had prevented them from conducting trout recovery activities in their forest. Officials from three other forests added that waterlogged or obstructed trails, which force visitors to create alternate routes around obstacles, have negative effects on the visitors, as well as on resources. For example, on one trail, OHV riders created trenches in a meadow to avoid water on the trail (see Figure 5), and, according to an agency official, at $100,000 per mile of trail, fixing the rutting by installing boardwalks to raise the trail above the surrounding meadow would be cost prohibitive. Another official gave an example of horseback riders' creating new stream crossings to avoid unsafe bridges. In addition to being potentially dangerous, such new crossings could damage resources by depositing additional sediment in creeks.

Source: Forest Service.

Figure 5. Visitor-Created Alternate Route to Avoid a Wet, Muddy Trail, Beaverhead-Deerlodge National Forest, Montana.

Source: GAO.

Figure 6. Deep Trenches Due to Inadequate Drainage, Gifford Pinchot National Forest, Washington, and Chattahoochee-Oconee National Forest, Georgia.

Delaying maintenance can also increase the effort required to perform routine maintenance and lead to increased maintenance costs in the future, as we have previously reported in other contexts.[16] Forest Service estimates of deferred maintenance needs include the one-time cost to conduct maintenance that has been deferred, but these estimates do not quantify the extent to which costs have increased over time as maintenance continues to be delayed. One forest official gave two examples of circumstances in which deferred maintenance could later increase costs—although the extent to which costs would increase depends on such factors as length of trail segment needing to be restored, distance from trailhead, and soil type—as follows:

- *Water-eroded trenches:* If drainage features such as water bars or drainage dips—which direct water away from trails to reduce erosion—are not regularly cleaned out, the drainage features can fail, and water can flow down the trail, creating deep trenches over time (see Figure 6). As a result, expensive maintenance is later needed to restore the trail in its existing location or to reroute it.

- *Inadequate trailside brush removal:* If brush alongside trails is not routinely removed, vegetation may grow and eventually take over the whole trail. Such overgrowth is especially common in areas of heavy rainfall, such as the Pacific Northwest and the Southeast, where, officials said, a trail can become overgrown in 5 years or less. Once a trail is overgrown, heavy maintenance is required to chop through roots and reestablish the trail's tread.

Officials from another forest told us that some trails in their forest are maintained so infrequently that by the time crews get to them, so much maintenance has been deferred that the trails need to be completely rebuilt. As one official said, "The longer one waits to fix a problem, the harder it will be to fix."

To Maintain Trails, the Forest Service Relies on a Combination of Internal and External Resources

The Forest Service relies on a combination of internal and external resources to help maintain its trail system. For example, the agency allocates some of its congressionally appropriated funds to support trail maintenance. In addition, the agency received about $100 million under the American Recovery and Reinvestment Act of 2009 for trail maintenance activities. External resources used by the agency for trail maintenance include volunteer labor and funding from federal programs, states, and other sources.

Internal Resources Include Funds Allocated to Trails and to Other Agency Programs

The Forest Service uses a variety of internal funding sources to support trail maintenance, according to officials we spoke with. The agency receives annual appropriations from Congress for capital improvements and maintenance, which it allocates to a variety of budget line items, including trails. This trails allocation is the agency's primary source of funding for trail maintenance activities. In fiscal years 2006 through 2012, the agency's annual trails allocation ranged from a low of about $73 million to a high of about $88 million, averaging about $80 million (see Figure 7).

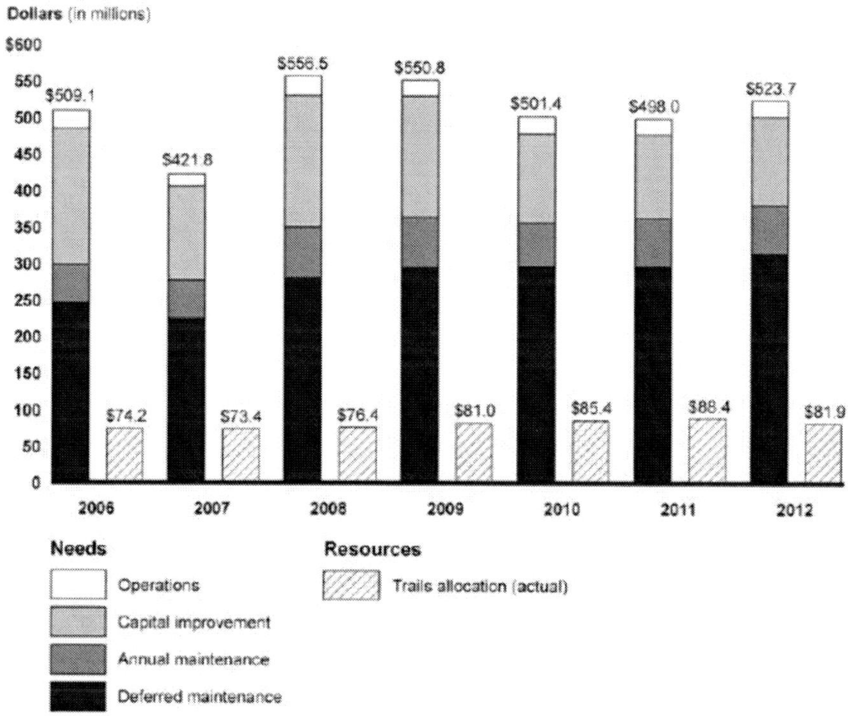

Source: GAO analysis of Forest Service data.

Figure 7. Estimated Trail Maintenance Needs, Compared with the Forest Service's Trails Allocation, Fiscal Years 2006 through 2012.

Not all of this money goes directly toward trail maintenance, however. As with other agency programs, a portion of the overall trail maintenance allocation is retained at the Forest Service headquarters level to cover agency overhead costs, before the remainder is distributed to the regions.[17] The regions likewise use a portion of the trails allocation to cover costs at the regional level before in turn distributing funds to individual forests for trail maintenance activities. For fiscal years 2010 through 2012, from 29 to 32 percent of the trails allocation was held at the national level for overhead costs. The regions also reported holding trails allocations at the regional level for purposes such as overhead costs, capital investment projects, and emergency reserves, before the remainder was distributed to forests.

Headquarters officials told us that since fiscal year 2007, they have used a historical model to determine how trails allocations should be distributed to each region. According to an agency budget official, the model evaluates three primary elements: the inventory of trails in the region, including trail miles and classes; status of the travel management planning process; [18] and the region's performance relative to agency priorities. For fiscal years 2011 through 2013, headquarters officials prorated and adjusted regional funding to meet national and region-specific needs identified by the agency's national and regional recreation directors, such as allocating funds to address an epidemic of mountain pine beetles in the Rocky Mountains. Regional portions of the Forest Service's trails allocation varied substantially; in fiscal year 2012, for example, after national cost pools were accounted for, regions received trails allocations ranging from $3.1 million to $9.7 million (see table 6 in app. III).[19]

After receiving their trails allocations, the regions in turn direct funding to national forests, and, regional officials told us, they take a variety of factors into account when doing so. As is done at the national level, six of the nine regions consider total number of trail miles, and one of these six also considers emerging issues, such as mitigation of mountain pine beetles, when determining annual allocations. Another region recently initiated a new process in which it gives a base administration amount of $60,000 to each forest, plus an additional amount tied to each "user visit" to the forest.[20] Officials from another region noted that their region's trails allocations to national forests are based on the amount of work forests can accomplish toward regional targets and extra trail needs, such as bridge replacements. Four of the nine regions noted that they hold back a portion of the trails allocation for capital investment projects related to trails. For example, one region funds one large trails capital investment project each year, valued at $125,000 to $250,000. According to regional officials, they established this practice to address high costs related to large capital investment projects, such as complex bridges, because a single large project could deplete a forest's entire trails allocation otherwise, and no other trail maintenance would be performed.

In addition to the Forest Service's trails allocation, the agency allocates funding to other programs that help support trail maintenance activities. 21 For example, officials from one forest reported that because trails staff also work for recreation programs, part of their salaries are paid from the national forest recreation and wilderness allocation, as well as from the trails allocation. Officials from this forest said interns and wilderness rangers funded through the national forest recreation and wilderness allocation do trails work in addition to interacting with visitors. The officials said that this practice has been very effective for addressing trail maintenance needs. Officials at other forests reported accomplishing trail maintenance through activities funded by the agency's integrated resource restoration allocation. This allocation was implemented on a pilot basis in certain regions in fiscal year 2012. Incorporating several existing allocations, the new allocation is intended to support actions to restore or sustain water quality and watershed processes, including road and trail restoration activities. Officials from some forests noted that because unmaintained trails may produce erosion adversely affecting water quality, they had used some of their integrated resource restoration allocation to conduct trail maintenance. Additionally, officials from a number of forests that had experienced wildland fires said they had used burned area emergency response allocations to address some trail maintenance needs on forests and rangelands affected by fires. These funds are available to support emergency response projects on lands damaged by wildfires.[22]

Source: GAO.

Figure 8. Bridge Replacement and Signs Paid for with Recovery Act Funding, Mt. Hood National Forest, Oregon.

In addition, the Forest Service allocated about $100 million of the funding it received under the American Recovery and Reinvestment Act of 2009 (Recovery Act) to trail maintenance and decommissioning activities, which some forest officials told us they used to help address their trail maintenance backlogs.[23] The amount of Recovery Act funds for trail maintenance and decommissioning distributed to the regions ranged from $540,000 to the Intermountain Region to over $19 million to the Pacific Southwest (see table 7 in app. III for information on Recovery Act funds allocated to regions and states). Of the 90 trail maintenance projects supported by Recovery Act funds, agency documents show that 76 addressed deferred maintenance, including 27 that repaired or replaced bridges. For example, Mt. Hood National Forest in Oregon received $1,400,000 to refurbish and repair trails to improve public access and hiker safety, which officials told us they used for a number of activities, including replacing 22 bridges and some signs (see Figure 8).

External Resources Include Volunteer Labor and Funding from Federal Programs, States, and Other Sources

In addition to internal resources, Forest Service officials reported using a number of external resources to support trail maintenance efforts, including volunteer labor and funding from other federal programs, states, and other sources. Volunteer labor is a particularly important resource for trail maintenance. In fiscal year 2012, the Forest Service reported that 1.2 million volunteer labor hours—or the equivalent of 667 full-time volunteers, valued at $26 million—directly supported its trail maintenance activities.[24] By comparison, in that same year, the Forest Service had the equivalent of 666 full-time trails employees.

The contributions of volunteers to trail maintenance may be higher than these figures indicate because volunteer hours may be underreported. According to agency documents, Forest Service staff are required to report the number of hours volunteers work on trails, but, according to an agency headquarters official, there are no annual agency targets for working with volunteers, and not all staff find the data valuable. Therefore, Forest Service staff may see little benefit in taking the time to collect and enter volunteer data, and, consequently, not all volunteer hours may be recorded. Moreover, some agency officials and stakeholders told us that not everyone who conducts maintenance on Forest Service trails is under a volunteer or challenge cost-share agreement, and informal contributions are not captured in the agency's volunteer data. For example, an official from one forest said that some visitors carry saws with them and remove deadfall or other vegetation they come across while using trails. These informal volunteer activities are not technically authorized or recorded in agency data, but an official from one forest said that forest officials "welcome the help."

Regarding external funds, all agency officials we interviewed at forests and ranger districts reported receiving external sources of funding from several sources, including other federal and state agencies. While the Forest Service tracks national grants and challenge cost-share agreements,[25] it does not centrally track external funding received by national forests and is unable to fully quantify how much total external funding the agency has received for trails. One key source of funding for trail maintenance is the Recreational Trails Program. Under this program, the Federal Highway Administration, in consultation with the Secretary of the Interior and the Secretary of Agriculture, makes funds available to states to award for trail maintenance or trail assessments. In fiscal year 2013, $80.2 million was set aside for this program nationally and was apportioned to the states. According to the officials we interviewed, states often grant a portion of these funds to national forests for trail maintenance or construction.[26] Officials from one forest we interviewed stated that they used funds from this source to install signs and reroute trails, and officials from another forest stated that they used the funds for major projects, including trail bridges. A third forest used $150,000 in Recreational Trails Program grant funding, combined with a grant from a local nonprofit, to pay for a professional trails assessment.

Officials from many forests we interviewed also told us they received state grants to support maintenance of trails for motor vehicles from their state's OHV program. Some states use funds collected from OHV registration fees to provide grants to local entities, including national forests, to maintain and improve trails for motorized users. Officials at one forest stated that the forest's ranger districts receive approximately $400,000 per year from their state's OHV registration fees, which the districts use to fund special projects, hire trails crews, and buy supplies to complete trail maintenance on Forest Service land. Officials at a ranger district stated that they received $239,000 per year in state OHV funding, which they used to fund a nine-person crew to maintain trails, among other activities. An official from this ranger district stated that much of the trail maintenance work funded by this grant was used to restore unauthorized routes that OHV users had created. Officials from another forest told us they receive $10,000 to $20,000 per year to maintain snow trails, plus an additional $10,000 to $20,000 per year to support OHV patrols, from their state's OHV program.

Officials from some forests we interviewed stated that they have also relied on funding from Title II of the Secure Rural Schools and Community Self-Determination Act of 2000 to conduct trail maintenance.[27] Under Title II of the Secure Rural Schools Act, projects may be funded for certain land management purposes that benefit federal lands, including projects related to the maintenance or obliteration of Forest Service roads, trails, and infrastructure.[28] Officials at one forest we interviewed stated that they had received from $18,950 to almost $97,000 in Title II funds each year and that their trail maintenance projects have relied heavily on this funding. Another forest reported receiving from $157,000 to $317,000 in Title II funding annually since 2009 for trail maintenance. These funds have allowed the forests to address some of their deferred maintenance backlog, as well as to complete annual maintenance. The authority to obligate funds for these projects is scheduled to expire in 2013, and officials at this forest stated that if they lost the funding, they would no longer be able to fund their seasonal trails crews and would be dependent on volunteers for needed maintenance, adding that some of their less-used trails would "go back to nature."

Officials from some forests told us they receive external funding for trail maintenance from other sources. For example, officials from one forest told us that the National Forest Foundation was raising funds to address trail maintenance issues in one area of their forest, [29] and that grant programs have been integral to their efforts to address deferred maintenance. They stated that in fiscal year 2012, the forest received nearly $200,000 in grants that they used to pay for a prison crew to maintain trails on the forest. Officials at a different forest told us that they received a $30,000 grant from an OHV manufacturer, using the funds to improve a trailhead and maintain OHV trails. Some forests also receive funding from trail user or outfitter fees that help fund trail maintenance. For example, officials at one forest told us that they collect about $30,000 to $40,000 annually in OHV user fees, some of which fund trail maintenance. Officials at another forest stated that they collect around $100,000 annually in outfitter and guide fees, which they use for various road and trail projects.

Resources Are Combined by National Forests in Different Ways

In our interviews with agency officials, including those at the national, regional, forest, and ranger district levels, we found that national forests and ranger districts combine funding and personnel resources in different ways to accomplish trail maintenance. Officials from a number of ranger districts told us that they rely on a combination of resources to maintain an effective trail maintenance program; as one regional official put it, the trail maintenance program "is held together by Band-Aids and baling wire." For example, a ranger district in one forest we visited used state grant dollars to pay for maintaining motor vehicle trails while volunteers conducted most maintenance on trails closed to motor vehicles. Officials from another forest told us that they use their trails allocation to pay for their basic trails program, including trails crew salaries and overhead costs, and grants and other external funding to pay for on-the-ground trail maintenance. An official in one district described his district's trails program as having a "large quiver of financial resources," which includes the trails allocation, state OHV grant funding, and partnerships with various organizations that contribute funding. Officials from this district also said that they benefit from a statewide trails crew that works on trails open to motor vehicles; the crew is paid for by the state's OHV program and works on motor vehicle trails on public lands throughout the state. Additionally, a number of forests we visited stated that they combined funding sources with volunteer or other labor sources to maintain their trails. For example, some forests have local groups who adopt trails or coordinate trail workdays, thereby taking responsibility for trail maintenance on one trail or trail segment. One ranger district we interviewed used its Secure Rural Schools Act Title II funding to pay for a trails crew on one side of the district, while relying entirely on volunteers on the other side. In another ranger district, officials reported that most of the maintenance of trails closed to motor vehicles is done by volunteers and that for heavy maintenance, such as tree removal, the district borrows a machine from another district. Some forests we visited are seeking new ways to complete trail maintenance. For example, officials from several of the forests and ranger districts we interviewed in Arizona, Colorado, and Idaho stated that they sometimes use prison crews because the crews are inexpensive and complete high-quality work. An official from one forest told us that although the forest must pay for the foreman and materials, it pays prisoners only $0.50 per day. As a result, it can generally accomplish maintenance work for 60 percent of what it would ordinarily cost to contract out the work, although an official noted that it takes forest officials more time to manage prison crew contracts than regular contracts.

Varied Factors Complicate the Maintenance of Forest Service Trails

According to agency officials and stakeholders we spoke with, a number of factors complicate the Forest Service's trail maintenance efforts, including (1) factors associated with the origin and location of trails, (2) some agency policies and procedures, and (3) factors associated with management of volunteers and other external resources. No single factor was identified as the most problematic; the types of factors identified, and the extent to which they complicate trail maintenance, varied across forests and regions.

Origin and Location of Many Forest Service Trails Require More Frequent and Resource-Intensive Maintenance Efforts

The origin of many system trails as legacy trails, roads converted to trails, or user-created trails, as well as the location of trails in designated wilderness or in areas affected by insect or disease outbreaks, wildland fire, or other natural events, complicate trail maintenance by requiring more frequent and resource-intensive trail maintenance efforts.

Trail Origin

Factors associated with the origin of many trails present a variety of complications in maintaining them, according to a number of agency officials and stakeholders we interviewed. Many Forest Service trails are legacy trails created for purposes other than recreation, such as access for mining, timber harvesting, or firefighting. Some of these trails were carved straight up steep slopes, leaving erosion-prone trails requiring continual maintenance; even on less-steep slopes, if a trail is built along a hill's fall line—the natural line down which water flows—it will naturally erode over time. Other trails were built through meadows, resulting in standing water on certain stretches, or in other problematic locations, such as on a stream bank (see Figure 9 for examples of these conditions). In addition, as part of the travel management process, many forests in recent years have converted Forest Service roads into trails open to motorized vehicles. Not all forests have been affected by these conversions, but officials from some forests said that conversion of hundreds or even thousands of miles of roads to motor vehicle trails had added new trail maintenance challenges and strained already-limited budgets.[30] Some officials told us they need heavier equipment and engineering expertise to address maintenance issues on many roads converted to trails; for example, as a result of one road-to-trail conversion, the trail system in one forest we visited had gained a two-lane car bridge across a wide river (see Figure 10).[31] Further, unauthorized trails created by users, which are not part of the agency's official trail system, take time and resources away from maintaining system trails because officials must address safety and resource concerns associated with the trails, according to officials we interviewed. Some officials told us their forests have hundreds of miles of user-created trails; in some areas, more of these trails exist than system trails.

Source: Forest Service.

Legacy trail built through a meadow led to standing water on the trail. A Forest Service Official told us that to protect the natural resource, boardwalks would need to be built through the meadow, but doing so would be cost prohibitive.

Figure 9. Legacy Trails in Chattahoochee-Oconee National Forest, Georgia, and Beaverhead-Deerlodge National Forest, Montana.

Source: GAO.

Figure 10. Road Bridge Adopted into Mt. Hood National Forest's Trail System, Oregon.

Many legacy and user-created trails are not sustainable over the long term, according to recent research and agency officials and stakeholders. These trails occupy terrain that is subject to severe erosion, require considerable ongoing maintenance, and do not meet users' needs without ecological damage. As a result, such trails require a disproportionate share of resources to maintain—akin to bandaging a wound that will never heal, in the words of one official. For example, one stakeholder told us about a Forest Service bridge to a waterfall, whose railing had been replaced 10-15 times in the past 20 years because the bridge was situated where, during severe weather, water would rush over a nearby cliff and rip out the handrail. The stakeholder commented that relocating the bridge would be more sustainable in the long term than continually repairing it. Similarly, officials from a Pacific Northwest forest told us that some of their forest's trails were built with major design flaws, such as trail segments where snow never melts. These officials said they have considered rerouting such sections to make them more sustainable, but doing so would require environmental review under the National Environmental Policy Act, which, they said, would be expensive; on the other hand, not going through this process contributes to the agency's backlog of deferred maintenance.[32] Many officials and stakeholders emphasized that despite the up-front costs of rerouting and reconstructing unsustainable trails, maintaining well-designed trails is much more cost-effective over the long term. For example, one official noted, the majority of the agency's trail maintenance costs are related to moving trails crews and equipment to the trails that need maintenance, and that well-designed trails cost less to maintain in the long term because crews do not have to visit them as often.

Trail Location

Maintaining trails located in wilderness areas or areas affected by insect or disease outbreaks, wildland fire, or other natural events can also be resource-intensive, in addition to being complicated by other factors. About 20 percent of the Forest Service's 158,000 miles of trails are located in designated wilderness areas, which can present challenges because of the prohibition against use of motorized equipment in designated wilderness and the remote location of many wilderness trails.[33] Regarding motorized equipment, officials and stakeholders hold different opinions about the legal prohibition against using chain saws and other motorized equipment in wilderness areas. Some officials and stakeholders told us this prohibition results in more time and effort to remove downed trees because maintenance crews must instead use nonmotorized tools, such as crosscut saws.[34] In certain situations, land managers may obtain an exemption from regional foresters allowing use of motorized tools to meet the minimum requirements for the protection and administration of a wilderness area, but requesting such an exemption is at the discretion of the local Forest Service manager, and some officials told us that certain managers are not inclined to request such exemptions, given the agency culture to preserve wilderness character.[35] In contrast, many officials and stakeholders we interviewed said that the general prohibition against power tools is not a complicating factor because crosscut saws are as efficient or nearly as efficient as chain saws, and chain saws are heavier to transport.[36] Several officials told us that accessing wilderness trails, often located deep in the backcountry, requires considerable time and effort. For example, officials from one forest said that it may take hours to drive to a wilderness trailhead, take 1 to 2 days to hike to the site needing maintenance, and require crews to stay overnight—adding to the cost and complexity of backcountry trail maintenance. The Forest Service's trail maintenance efforts are also complicated when trails are located in areas affected by insect or disease outbreaks, wildland fire, and other natural events. National forests in some western states have suffered heavily from a mountain pine beetle epidemic, which has left many dead or dying trees that are starting to fall, sometimes across or near trails. Officials from one forest told us their forest's entire trails program does little beyond removing hazardous trees because beetles have killed so many trees. Officials in other parts of the country told us that their trail maintenance programs were being affected by other insects, such as the hemlock woolly adelgid, or by diseases, such as laminated root rot in Douglas-fir trees. 37 Wildland fire also complicates trail maintenance. According to officials, a number of steps may be needed before a trail can be reopened after a wildland fire, such as removing hazardous trees, relocating drainage features, and stabilizing rocks. In addition, a number of forest officials told us that other natural events, such as tornadoes, hurricanes, floods, and windstorms, sometimes complicated their trail maintenance. For example, in the Pacific Northwest, officials from two forests told us that storms may cause flooding and landslides that easily wash out trails because of the region's loose volcanic soils.

Various Agency Policies and Procedures Can Limit Agency Trails Expertise and Take Time Away from Trail Maintenance on the Ground

Additional factors complicating the Forest Service's trail maintenance activities include the absence of a career path or training program for trails staff, which can limit agency expertise; burdensome data collection efforts; and certain administrative procedures that take time away from conducting maintenance on the ground.

- *Career path, training.* Many officials noted that the Forest Service has no career path or training programs for trails staff, which makes it difficult for the agency to develop and retain professional expertise and leadership for the trails program. For example, because full-time, permanent trails positions do not always exist at the district or forest levels, the agency often hires temporary or permanent-seasonal employees to maintain trails.[38] These employees, however, often work for only one or two summers, requiring local officials to hire and train new trails employees the following season. Several officials and stakeholders told us that because of retirements and attrition, the agency has lost almost all of its trails expertise in recent years, and other officials noted that certain technical skills—such as using crosscut saws, working with horses, or blasting rock—are becoming more difficult to find when seeking new trails employees. The Forest Service currently has no national, standardized training for these skills. (Staff training, retention, and expertise are discussed in more detail later in this report.)
- *Collecting trail condition data.* Many local trail managers told us that the effort needed to collect trail condition data each year is burdensome and takes time away from conducting on-the-ground trail maintenance—an important consideration given the limited resources available to them. Many also said they do not use the collected information for making decisions, such as setting priorities, at the local level and use it only for upward reporting. Agency headquarters officials, however, emphasized to us the importance of data collection for estimating trail maintenance costs nationwide, as well as for providing information on trail conditions to local officials.
- *Administrative procedures.* Officials and stakeholders also identified a number of administrative and other factors that complicate trail maintenance, some of which are outside of the agency's control:
  - *Efforts to reduce travel costs.* Many officials said that agency efforts to reduce travel costs have hindered their ability to complete trail maintenance on the ground, especially on remote trails. Several officials told us that trails crews who in the past may have been allowed to spend the night near a work site must

now travel back and forth each day to avoid food or lodging costs. As a result, more time is spent transporting crews—up to several hours each way—and less time is spent completing work on the ground.

- *Environmental review processes.* Other officials and stakeholders said that analyses required under the National Environmental Policy Act can be expensive and time-consuming, thereby detracting from actual maintenance activities. Routine trail maintenance does not require detailed environmental analysis, but the agency sometimes performs such an analysis for new trail construction, trail relocations, and other substantial trail work.
- *Budget timing.* The Forest Service does not always have a final budget in place for a given fiscal year until spring, which some officials said affects their ability to plan and execute trail maintenance. For example, one official said, they cannot sign and execute contracts until they have an approved budget, which may happen late in the fiscal year when contractors are already committed to other projects. Also, officials from one forest told us that because of their forest's high elevation and persistent snowpack, they can work only during a 6-to-8-week window in late summer. Timing of the budget, along with a short season, can make it hard to complete trail maintenance.

Managing Volunteers and Other External Resources Can Take Time Away from On-the-Ground Maintenance

Although volunteers and other external resources were repeatedly cited as important to the agency's trail maintenance efforts, officials and stakeholders we interviewed identified a number of complications related to working with volunteers, including insufficient agency emphasis on managing volunteers; the time and effort it takes to coordinate, train, and supervise them, which decreases the time officials can spend conducting maintenance; safety and liability concerns that limit local use of volunteers; and the tenuous nature of partnerships. In addition, officials noted that managing other external resources for trail maintenance, such as time required to research and apply for grants, can detract from performing maintenance on the ground.

- *Emphasis on volunteers.* According to some agency officials and stakeholders, the Forest Service recognizes but does not always sufficiently emphasize managing volunteers when it hires and trains trails employees. Congress and the executive branch, including the Forest Service, have recognized the importance of volunteers to complement the agency's work in trail maintenance and other activities. For example, Executive Order 13195, issued in 2001, directs agencies to engage volunteers in all aspects of trail planning, development, maintenance, management, and education, as outlined in the National Trails System Act. The Forest Service has also emphasized the importance of volunteers in the chapter on volunteer management in the *Forest Service Manual*. Even so, at the forest and district levels, volunteer management is generally a collateral duty, and collaboration with and management of volunteers are not clear expectations of trails staff. One official pointed out that it takes the "right type of Forest Service employee to build partnerships," stating that the agency should be more diligent in hiring trails coordinators with collaboration skills. Moreover, some officials and stakeholders pointed out that the Forest Service provides limited training to staff who manage volunteers. For example, one official noted, the agency conducts quarterly web-based workshops on working with volunteers but offers little additional training to field staff who work with volunteers. (Volunteer management is discussed in more detail later in this report.)
- *Coordinating, training, and supervising volunteers.* Many Forest Service officials told us, and we have previously found, that coordinating, training, and supervising volunteers take effort, as well as time away from other tasks;[39] in the words of many officials we spoke with, "Volunteers aren't free." Officials from the majority of forests we visited told us that they did not have sufficient staff or resources to effectively manage additional volunteers; three forests reported turning away volunteers as a result. In contrast, officials from other forests we visited told us that they never turned away volunteers and had the capacity to manage more volunteers, particularly when groups are skilled and can perform maintenance on their own. On the other hand, some groups are not capable of operating without supervision; several officials said that undirected or unsupervised volunteers or youth crews may damage trails and that Forest Service crews sometimes have to revisit volunteer-maintained trails to repair

volunteer-caused damage or complete maintenance not done to Forest Service standards.

- *Safety or liability concerns.* Officials and stakeholders also told us that factors related to safety and liability sometimes complicate working with volunteers. For example, some forests do not allow volunteers to use chain saws, while other forests vary in their certification requirements for volunteers to use equipment such as crosscut saws or chain saws. Officials and stakeholders told us that some forests require a 40-hour training session to use chain saws, while other forests require a 1-day or weekend course. Moreover, some but not all forests accept saw certifications awarded by other forests. Many officials told us that safety is a top priority, and managers are sometimes hesitant to allow volunteers to use equipment if they risk being hurt and filing a workers' compensation claim. Volunteers are considered federal employees under the Volunteers in the Forest Act for tort or workers' compensation claims. Since workers' compensation is generally covered by local units, one claim may consume a local unit's entire annual trails allocation, according to some officials and stakeholders.
- *Tenuous nature of partnerships.* Some officials told us that relationships with partners can be tenuous, which can make volunteers less willing to work with the agency. In some cases, volunteer groups will support the Forest Service as long as the agency is supporting their values but can turn into adversaries if the agency makes a decision they do not agree with—for example, if, to protect natural resources, the agency closes a trail volunteers like. Volunteers also may develop a sense of trail ownership. Such pride of ownership may confer an advantage as volunteers try to do a good job maintaining trails, but, according to officials, it can also present challenges when volunteer groups want to influence agency decisions about trail maintenance priorities.
- *Applying for and managing external funding.* Officials we interviewed also observed that, as in working with volunteers, it takes time to apply for external funding and manage requirements associated with this funding, which allows less time for actual trail maintenance. For example, officials told us, it takes time and effort to seek and apply for external grant funding and to meet requirements for such outside funding once received. Officials from one forest said they could not at the time manage additional grants because they did not have the time

or staff, and officials from other forests said it is hard to keep up with reporting or other administrative requirements for trails projects funded with external resources. Nevertheless, some officials told us that even with the additional effort needed to comply with these requirements, external funding is critical to their trail maintenance efforts.

STAKEHOLDERS HAVE IDENTIFIED NUMEROUS OPTIONS FOR IMPROVING TRAIL MAINTENANCE OVER THE LONG AND SHORT TERMS

Agency officials and stakeholders identified numerous options aimed at improving Forest Service trail maintenance, which generally fell into the following categories: (1) assessing the sustainability of the trail system, (2) improving certain policies and procedures associated with the Forest Service's management of the trails program, and (3) better using volunteers and other external resources.

Assessing the Sustainability of the Trail System

Many officials at all levels of the agency, as well as some stakeholders we met with, stated that the Forest Service's trails program might benefit if the agency were to systematically assess its trail system. In 2010, the Forest Service issued a document titled *A Framework for Sustainable Recreation*, in which the agency presented a strategic vision and guiding principles to achieve sustainability in all aspects of its recreation program, including trails. As part of this vision, the *Framework* noted the importance of the Forest Service's evaluating its infrastructure investments and program costs to identify "the gap between program needs and available resources . . . along with options for closing the gap." Many officials and stakeholders we interviewed told us that trail systems should be "right-sized"; that is, units should assess their trail inventories in light of the resources available for maintenance and take steps, such as closing trails or portions of trails or reducing the maintenance on certain trails, so as to narrow the gap between funding and maintenance needs consistent with the *Framework*. One approach that the Forest Service used in the mid-2000s is the agency's recreation facilities analysis process, which assessed recreation sites—such as campgrounds, day-use sites, and some trailheads (but not trails themselves)—to "assist forests in creating a sustainable program that aligns recreation sites with visitors' desires, expectations, and use." This process resulted in relatively few decommissioned sites,[40] but a headquarters official told us the process benefited local units by helping them identify a variety of tools to address the gap between program needs and available resources. Similar to this past approach, the Forest Service is developing a process intended to help forests achieve a trail system that meets community needs, does not harm natural resources, and can be maintained with available resources. Headquarters officials told us the agency had not yet determined how it will implement the process or the time frames for doing so.

Trail assessments could also improve the physical sustainability of individual trails. Numerous stakeholders and officials noted the importance of sustainability in the trail system, stating that redesigning legacy trails and relocating unsustainable trail segments—through rerouting steep segments to reduce erosion, for example—would substantially reduce maintenance work over the long term. Some of these officials and stakeholders acknowledged the potential for considerable up-front costs to relocate unsustainable trail segments but stated that long-term maintenance costs would be significantly lower for well-designed trails. One Interior agency official said that a potential strategy would be to address unsustainable trails in feasible portions by undertaking trail reroutes and redesigns on a certain percentage of the trail system each year. For example, by annually addressing 5 percent of the system, the agency would have "solved its trail problems" within 20 years, according to this official, and be better positioned to address needed yearly trail maintenance.

Some officials indicated that training Forest Service employees on sustainable trail design might also improve trail sustainability, noting that agency field staff may not have a full understanding of how to assess trails for sustainability or how to redesign or relocate unsustainable trails because the agency has provided little guidance or training on this. As noted, the agency does not have a robust trails training program, and while the concept of trail sustainability is discussed in some of the agency's guidance on trail design, little hands-on training is provided to show field staff how to implement this guidance on the ground.

Although the Forest Service offers little training on assessing sustainability, some forests we visited had already taken steps to assess the sustainability of their trails and to identify and implement opportunities to reroute or otherwise improve them, consistent with the *Framework*. For example, one forest surveyed 250 miles of trail and is analyzing data from its assessment to identify unsustainable trails and set priorities for work, including identifying trails to add, decommission, or reroute. Other forests we visited were taking other approaches toward more sustainable trails. For example, one forest assessed its road and trail systems together— rather than focusing on just roads and trails used by motorized vehicles— as part of its travel management planning. The forest has undertaken a separate analysis to look at the efficiency of its current approach to managing and maintaining its trail system.

Another forest we visited was conducting systematic assessments on particular trails or trail systems throughout the forest; officials told us they had decided not to spend money on unsustainable trails and were actively relocating these trails. This forest had assessed all of its OHV trails, for example, and, on the basis of this assessment, had repaired and rerouted certain trails and implemented seasonal and weather-related closures.

On the other hand, not all forests have assessed the sustainability of their trails or identified opportunities for improvement, and because such assessments—and subsequent changes to trail systems—can be costly, time-consuming, and contentious, the agency has not undertaken or promoted such assessments nationwide. Without doing so, however, the agency may continue to devote substantial resources to maintaining inadequately designed trails. For example, officials we spoke with at one forest were in the process of rebuilding trails destroyed by a fire and told us that rebuilding the approximately 300 miles of trail affected by the fire would cost almost $750,000. They had not, however, assessed the sustainability of those trails to determine the extent to which rerouting unsustainable trail sections now would save the agency funding and resources later.

Improving Forest Service Policies and Procedures

Some officials and stakeholders also identified a number of options related to improving Forest Service policies and procedures to better manage the trails program, including the following:

- *Implement standardized trails training.* Some officials and stakeholders stated that the agency would benefit from a training curriculum about basic trail design, construction, and maintenance—

to go beyond the sustainability training noted earlier—aimed at providing basic field skills to staff responsible for trails. The agency does not have a robust trails training program, and a number of officials and stakeholders said that training was needed on basic field skills. The *Framework* states that the agency is to train staff and develop needed skills. Agency headquarters officials agreed that training is important and would be best conducted in the field, but they noted that because providing in-person training in the field is expensive, the agency has shifted heavily to web-based training. Nevertheless, given the nature of trail maintenance work, some officials emphasized to us the importance of conducting such training in person. Without in-person training, agency staff may not have the skills they need to perform on-the-ground trail maintenance activities.

- *Improve expertise by recruiting and retaining skilled trails employees.* The *Framework* calls for the agency to improve its expertise by recruiting and retaining staff with needed skills. As noted earlier, however, the agency has had difficulty hiring and retaining skilled trails employees. Many officials stated that taking steps to hire and retain skilled trails employees would improve trail maintenance; this option was of particular interest to a number of regional representatives we spoke with. For example, officials from one forest said, revised job descriptions might help recruit trails employees who are more knowledgeable about trail maintenance and management. Further, according to some officials, if the agency could create incentives for skilled trails employees, such as hiring them at higher pay or having greater opportunities for promotion, they might be more likely to stay in trails positions, and the agency could retain their expertise. Other officials said that it would also be helpful if the agency's hiring policies made it easier to move temporary workers into permanent positions. Without policies and practices that promote hiring and retention of skilled employees, the agency cannot ensure that it has the needed expertise to maintain trails.
- *Improve data collection practices.* Many officials told us that the agency could streamline or otherwise improve practices for collecting trails condition data to make the process less burdensome and the data more useful. Agency officials, acknowledging that the surveys are time-consuming, said they are pursuing an initiative to streamline how the data are collected—an initiative that has been under way since 2006. Specifically, one official told us the agency intends to replace

the current system—which requires staff to fill out paper surveys while on the trail and then manually enter the information into an agency database—with a process for electronic field data collection that relies on handheld tablet computers, synchronized with a wireless distance-measuring device, which automatically upload collected data to the database. Officials told us they hope to introduce the new process in 2013 or 2014.

- *Assess how the agency distributes trails funding.* Some officials told us that the agency may benefit from changing the way it distributes its trails allocation funds to regions and forests. Some officials told us that trails allocation funding should be linked to the number of visitors forests receive. Other officials disagreed, however, noting that the agency's multiple-use mission is to accommodate different recreational experiences, including solitude and a wilderness experience on little-used trails. Moreover, some expressed concern about the reliability of agency data on visitor use and relying on these data as the basis for distributing funds. The trails program currently has a working group composed of regional trail coordinators who are evaluating the national process for distributing trails allocation funds and potential alternatives, including reviewing existing distribution models used by regions to see if any might be applicable at the national level. A headquarters official noted that, since each region experiences different circumstances, the exercise has been difficult because the working group is finding that one model does not necessarily fit the needs of the entire country. This official added that it is not clear when or if a new model will be applied to distribute trail maintenance funding but said that the agency is aiming to implement a new process in fiscal year 2014.
- *Improve the sharing of best practices across the agency.* Some officials and stakeholders told us that the agency could improve how it shares best practices or success stories related to trail maintenance across the agency. For example, officials from one forest said they had few opportunities to share with other forests what they had learned over the last few years about designing sustainable trails.

Improving Management of Volunteers and Other External Resources

Recognizing the considerable time volunteers donate to trail maintenance efforts, some officials and stakeholders stated that improving management of volunteers would make working with them easier and more effective. This option is consistent with the vision the agency has presented in the *Forest Service Manual*, which articulates the agency's goal to recruit, train, and use the services of volunteers to complement its trail maintenance and other work. Officials and stakeholders identified a number of ways to enhance the agency's use of volunteers and partnerships, including the following:

- *Make volunteer and partnership management a clear expectation for trails staff, and increase training.* As noted earlier, even with the agency's emphasis on using volunteers—articulated in the *Forest Service Manual*—the agency has not established collaboration with and management of volunteers as clear expectations for trails staff responsible for working with volunteers, and training in this area is limited. Given the value of volunteer hours devoted to Forest Service trail maintenance in fiscal year 2012—equivalent to nearly one-third of the agency's trails allocation—some officials and stakeholders said that making collaboration with and management of volunteers clear expectations for trails staff (e.g., through performance evaluation standards) and offering relevant training could enhance the agency's management of volunteers, as well as better reflect the central role that volunteers play in trail maintenance. Other officials said that the agency should consider hiring dedicated volunteer coordinators at the forest and ranger district levels. A headquarters official told us that the Forest Service has been slow to update its policies and practices to reflect its increasing reliance on volunteers, in part because the agency has not made it a priority—as evidenced by the agency's treatment of volunteer management as a collateral duty.

  Nevertheless, without making collaboration with and management of volunteers a clear expectation for trails staff who work with volunteers and offering relevant training, the agency cannot be sure it is fully capitalizing on the assistance volunteers can offer.
- *Improve consistency of volunteer management policies, including certifications.* A number of officials and stakeholders said that making agency policies, regulations, and certification processes more uniform

would make it easier for people to volunteer for the agency. Several brought up the issue of inconsistent saw certification requirements across districts and forests, stating that having consistent procedures for certifying volunteers would make it easier for volunteers to help maintain trails in more than one forest. To address this issue, the agency is developing a proposed directive to provide national guidance for training and certification in saw use, which would apply to both crosscut saws and chain saws. A headquarters official said that the agency originally intended to have the new saw directive finalized by summer of 2013 but that it now planned to seek public comment on the proposed directive in fall 2013 before it is made final.

- *Address liability concerns.* A number of officials and stakeholders said that changing how the agency handles workers' compensation claims may increase local volunteer participation. To overcome local officials' reluctance to use volunteers for fear that a workers' compensation claim might consume their entire trails allocation, some officials suggested that having a national funding source to pay workers' compensation claims would make local managers more willing to use volunteers for trail maintenance. A headquarters official told us that the Forest Service had explored moving to a national funding source in the past but had rejected the possibility because of the agency's interest in diverting less funding to cost pools overall. In addition, some officials and stakeholders said that changing how liability is handled in challenge cost-share agreements—under which liability generally rests with partner organizations—might increase volunteer participation if the agency were to assume this liability, because more organizations would be willing to volunteer under these agreements. A headquarters official said that the Forest Service is considering such changes, which would potentially require new legislation, so that the agency could take on liability for volunteers under both volunteer and challenge cost-share agreements. In addition, officials told us, the agency is preparing guidance on using both challenge cost-share and volunteer agreements simultaneously to address liability concerns in certain situations.

Some agency officials and stakeholders also identified ways they believe the Forest Service could better leverage external funds. For example, some officials and stakeholders said, forests could seek more grants to be used for trail maintenance, and officials from one forest said that units might benefit from hiring full-time grant administrators, who could help identify and administer available grants. Other officials said it would be helpful to have a headquarters official coordinate and share grant opportunities and new funding sources with field units. Headquarters officials acknowledged they could improve how they coordinate and distribute information on available funding to the field, but they also told us that the agency does not have enough staff to dedicate someone to looking for and informing field units of grant opportunities.

CONCLUSION

Charged with managing and maintaining some 158,000 miles of trails across the National Forest System, the Forest Service largely succeeds in offering trail users recreational opportunities ranging from solitary wilderness hikes to OHV access. The agency continually brings together personnel, equipment, and funding from numerous different internal and external resources to maintain trails—and indeed, the forests we visited were generally able to maintain their most popular trails and address safety concerns. Nevertheless, maintenance issues abound, and given the magnitude of the trail system, including many unsustainable trails, and limited available resources, the agency is facing a maintenance problem it is unlikely to completely resolve. Without conducting an analysis of trails program needs and available resources, consistent with the agency's *Framework for Sustainable Recreation*, and assessing potential ways to narrow the gap between them, the agency is likely to continue operating in a reactive mode, addressing short-term maintenance needs without a long-term understanding of how to better address the issue. The agency has recognized the importance of trail sustainability in reducing needed maintenance—for example, through its *Framework*—but it has not yet translated this emphasis into action in the form of training on sustainable design or local assessments that reevaluate both the uses of trails and their physical condition with longterm sustainability in mind. Even with such steps toward sustainability in the long term, however, certain agency policies and procedures may still make it difficult to keep up with maintenance needs in the short term. For example, the cumbersome approach to collecting and recording trail condition data, which the agency has been trying to streamline through electronic data collection since 2006, can hinder maintenance activities. Further, without policies that help in recruitment and retention of skilled employees—or basic, standardized in-person employee training on trail skills and on-the-ground maintenance—the agency may find itself without sufficient expertise to conduct needed maintenance. Furthermore, even with its extensive reliance on volunteers and the vision set forth in the *Forest Service Manual*, the agency continues to assign its employees volunteer management as a collateral duty and has not made collaboration with and management of volunteers clear expectations of trails staff or offered substantial relevant training. Without short- as well as long-term steps to adjust and streamline such policies and procedures, the Forest Service is likely to continue falling behind in maintaining its trails, spending scarce resources on unsustainable trails and presiding over degraded visitor experiences and natural resources.

Recommendations for Executive Action

To enhance the overall sustainability of the Forest Service's trail system, consistent with the vision articulated in *A Framework for Sustainable Recreation*, and to reduce the trail maintenance backlog, we recommend that the Secretary of Agriculture direct the Chief of the Forest Service to take several actions to improve the agency's trail maintenance approach in both the short and long terms.

To improve agency management of its trails program in the long term, particularly in light of the gap between program needs and available resources, the agency should take the following two actions consistent with the agency's *Framework for Sustainable Recreation*:

- In line with the *Framework*'s emphasis on evaluating infrastructure investments and program costs, (1) ensure that the agency's management of its trails program includes an analysis of trails program needs and available resources and (2) develop options for narrowing the gap between program needs and resources.
- In line with the *Framework*'s emphasis on sustainability, and to enhance trail sustainability over the long term, (1) improve guidance and increase training on sustainable trail design and (2) when appropriate, begin systematic, unit-level trail assessments that reevaluate trails with long-term sustainability as a goal.

To improve the agency's ability to keep up with its maintenance goals in the short term and reduce its maintenance backlog, the agency should take the following two actions:

- Take steps to improve policies and procedures related to trail maintenance. Such steps should include implementing electronic collection of trail condition data and offering more standardized in-person training on trail skills and on-the-ground maintenance. They could also include, for example, changing policies and practices to improve recruitment and retention of employees with trail expertise.
- Recognizing the importance of volunteers for trail maintenance, take steps to improve management of volunteers, including by ensuring that collaboration with and management of volunteers are clear expectations of trails staff and offering relevant training.

AGENCY COMMENTS AND OUR EVALUATION

We provided a copy of this report for review and comment to the Department of Agriculture. In written comments responding on behalf of the Department of Agriculture, the Forest Service generally agreed with our findings and recommendations. The Forest Service emphasized its commitment to implementing its *Framework for Sustainable Recreation*, including improved guidance and training on sustainable trail design. It also stated its commitment to improving policies and procedures related to trail maintenance, including implementing electronic collection of trail condition data, exploring options to improve recruitment and retention of employees with trails expertise, and improving collaboration with and management of volunteers. The agency noted, however, that its ability to take action in some of these areas, such as providing in-person training on trails skills, may be limited by budgetary constraints. The Forest Service also provided technical comments, which we incorporated as appropriate.

Anne-Marie Fennell
Director, Natural Resources and Environment

APPENDIX I: OBJECTIVES, SCOPE, AND METHODOLOGY

Our objectives were to examine (1) the extent to which the Forest Service is meeting trail maintenance needs, and effects associated with any maintenance not done; (2) resources, including funding and labor, that the agency employs to maintain its trails; (3) factors, if any, complicating agency efforts to maintain its trails; and (4) options, if any, that could improve the agency's trail maintenance efforts.

To conduct this work, we reviewed relevant laws and agency documents, including agency handbooks and other guidance. We interviewed Forest Service officials in headquarters and received information from all nine regions about trail maintenance needs and effects associated with any deferred maintenance. We also interviewed officials from a nonprobability sample of 18 national forests located in five of the nine Forest Service regions; we visited 16 of these forests and interviewed officials from 2 more. (Table 2 shows the forests included in our review.) During these visits, we held semistructured interviews with officials to learn about their trail maintenance programs; we also examined trails on which maintenance had been deferred, as well as trails that were well maintained. We selected these forests to represent variation in geography, proximity of forests to urban and rural areas, trail mileage, and type and intensity of trail use, although findings from this selection of forests are not generalizable to the entire population of national forests. We obtained data on the Forest Service's trail inventory for fiscal years 2008 to 2012 from the agency's Infrastructure database (known as Infra). To assess the reliability of the data, we reviewed relevant documentation and interviewed agency officials knowledgeable about the data. We determined that these data were sufficiently reliable for the purposes of this report.

To evaluate the resources the Forest Service employs to maintain its trails, we reviewed agency budget documents for fiscal years 2006 to 2012. We also collected and reviewed evidence from national, regional, forest, and ranger district officials about how funds are allocated for trail maintenance activities. In addition, we examined the agency's use of external resources in conducting trail maintenance and also the laws, regulations, and agency guidance governing the Forest Service's authority to use these resources. During our visits to national forests, we discussed and reviewed documentation related to their use of external funds for trail maintenance. We also interviewed an official from the U.S. Department of Transportation's Federal Highway Administration to learn more about the Recreational Trails Program, as well as an official from the Colorado Department of Natural Resources' Parks and

Wildlife division to learn about the state's grants program for trails used by OHVs. To evaluate the extent to which volunteers maintain trails, we reviewed agency volunteer data available for the most recent fiscal years, 2011 and 2012. To assess the reliability of the data, we reviewed relevant documentation and interviewed agency officials knowledgeable about the data; we found these data to be sufficiently reliable for the purposes of this report. We also interviewed headquarters officials to discuss volunteer management policies and officials at regions and forests to discuss the benefits and drawbacks of using volunteers to maintain trails. We also conducted semistructured interviews with representatives from a nonprobability sample of 16 nongovernmental organizations about their organizations' efforts to help the Forest Service maintain trails and about their views on Forest Service trail conditions. We selected these organizations to represent a variety of trail user, conservation, and industry perspectives. The views of representatives from these organizations are not generalizable to other nongovernmental organizations, but they provided various perspectives on the Forest Service's trail maintenance efforts. (Table 3 lists the organizations we interviewed.)

Table 2. National Forests Included in GAO's Review

Region number	Region name	National forest	State
1	Northern	Beaverhead-Deerlodge National Forest	Montana
		Bitterroot National Forest	Idaho and Montana
		Gallatin National Forest	Montana
		Idaho Panhandle National Forests	Idaho
		Nez Perce-Clearwater National Forests	Idaho
2	Rocky Mountain	Arapaho and Roosevelt National Forests	Colorado
		Medicine Bow-Routt National Forests	Colorado and Wyoming
		Pike and San Isabel National Forests	Colorado
3	Southwestern	Apache-Sitgreaves National Forests	Arizona
		Cibola National Forest	New Mexico
		Coronado National Forest	Arizona
		Gila National Forest	New Mexico
6	Pacific Northwest	Gifford Pinchot National Forest	Washington
		Mt. Hood National Forest	Oregon
		Willamette National Forest	Oregon
8	Southern	Chattahoochee-Oconee National Forest	Georgia
		Cherokee National Forest	Tennessee
		Francis Marion and Sumter National Forests	South Carolina

Source: GAO.

Table 3. Organizations GAO Interviewed

Organization	Geographic coverage
Back Country Horsemen of America	National
BlueRibbon Coalition	National
Friends of Northern Arizona Forests	Regional
International Mountain Bicycling Association	International
National Association of Forest Service Retirees	National
National Off-Highway Vehicle Conservation Council	National
Oregon Equestrian Trails	Regional
Pacific Crest Trail Association	Regional
Panhandle Trail Riders Association	Regional
The Partnership for the National Trails System	National
Professionals for Managed Recreation	National
Professional Trailbuilders Association	National
Southern Off-Road Bicycle Association	Regional
Washington Trails Association	Regional
Western Environmental Law Center	Regional
The Wilderness Society	National

Source: GAO.

To obtain information on any factors complicating trail maintenance and what options, if any, could improve it, we asked agency officials at all levels about both topics. Further, we convened a structured discussion group to gather perspectives from knowledgeable Forest Service officials representing all nine regions regarding challenges to maintaining trails and options for improving trail maintenance. We convened the discussion group via conference call and used web-based software to compile participants' comments. In our interviews with nongovernmental organizations, we asked for their views on challenges faced by the agency in performing trail maintenance and their views on any options for improvement. We also interviewed officials from three other federal land management agencies—the Department of the Interior's Bureau of Land Management, Fish and Wildlife Service, and National Park Service—to learn about these agencies' trail maintenance programs. We interviewed an official from Interior's U.S. Geological Survey to learn about current research on trail design.

We conducted this performance audit from June 2012 to June 2013 in accordance with generally accepted government auditing standards. Those standards require that we plan and perform the audit to obtain sufficient, appropriate evidence to provide a reasonable basis for our findings and conclusions based on our audit objectives. We believe that the evidence obtained provides a reasonable basis for our findings and conclusions based on our audit objectives.

Appendix II: Selected Trails Data from the Forest Service's INFRA Database

Table 4. Forest Service's National Forest System Trail Inventory and Estimated Number of Visitors, as of May 2013

Region numbera	Region	Total trail miles	Wilderness miles	Miles open to motorized vehicles	Miles closed to motorized vehicles	Estimatednumber ofvisitorsb
1	Northern	28,151	5,466	12,405	15,746	9,378,000
2	Rocky Mountain	19,011	4,549	7,035	11,976	27,962,000
3	Southwestern	9,993	3,398	1,324	8,669	18,185,000
4	Intermountain	29,348	5,233	13,370	15,978	22,422,000
5	Pacific Southwest	15,374	4,997	4,689	10,685	26,673,000
6	Pacific Northwest	25,326	6,105	10,338	14,988	15,428,000
8	Southern	11,133	904	1,908	9,225	28,993,000
9	Eastern	16,737	1,268	7,632	9,105	15,701,000
10	Alaska	1,439	92	577	862	1,138,000
Total		156,512c	32,014	59,278	97,234	165,880,000

Source: Forest Service.

Note: Numbers may not total because of rounding.

[a]The Forest Service does not have a region 7.

[b]This information, which comes from the agency's National Visitor Use Monitoring program, refers to the overall number of visitors to forests in each region, and encompasses all recreational activities on those lands, not solely trail use. The program is completed in 5-year cycles; the annual estimate for the number of visitors in fiscal year 2012 is based on visitation data collected from fiscal years 2007 through 2011.

[c]For purposes of this table, we obtained the most current information available during our review, that is, data as of May 2013. Because of routine fluctuations in trail inventory, however, total trail miles as of May 2013 (156,512) differ from total trail miles as of the close of fiscal year 2012 (158,104) discussed elsewhere in this report.

The tables in this appendix provide information on the Forest Service's National Forest System trail inventory from the agency's Infra database. Table 4 shows, for each region, total trail miles, wilderness miles, miles open to motorized vehicles, and miles closed to motorized vehicles. It also provides estimates of annual visitors to each region. Table 5 shows trail miles by trail class for each region.

Table 5. Forest Service's National Forest System Trail Inventory, by Region and Trail Class, Fiscal Year 2012

Region number[a]	Region name	Trail classes 1	2	3	4	5	Total trail miles[b]
1	Northern	1,285	9,918	12,185	5,264	15	29,039
2	Rocky Mountain	950	6,393	9,510	1,261	50	18,769
3	Southwestern	1,803	4,926	4,056	483	19	11,469
4	Intermountain	2,067	9,686	12,954	3,433	35	28,512
5	Pacific Southwest	472	3,872	8,818	1,729	65	15,349
6	Pacific Northwest	715	5,401	16,029	2,590	75	25,447
8	Southern	347	1,698	8,012	1,177	16	11,261
9	Eastern	689	3,514	10,287	2,319	29	16,850
10	Alaska	115	380	860	41	8	1,408
Total miles in trail classes		8,443	45,787	82,711	18,297	312	158,104
Total percentage in trail classes[c]		5.3	29.0	52.3	11.6	0.2	

Source: Forest Service.

Note: Numbers may not total because of rounding. [a]The Forest Service does not have a region 7.

[b]Total trail miles include trails for which the trail class has not been specified, a total of 2,554 miles.

[c]The total percentage in trail classes excludes trails for which trail class is not specified (2,554 miles, or 1.6 percent of trails) and therefore does not total 100 percent.

APPENDIX III: FOREST SERVICE TRAILS-RELATED BUDGET DATA

The tables in this appendix provide information on the Forest Service's trails allocations. Table 6 provides trails allocation data by region for fiscal years 2006 to 2012. Table 7 describes American Recovery and Reinvestment Act funding, by region and state, to support trail maintenance and decommissioning projects.

Table 6. Forest Service Trails Allocations Distributed to the Regions, Fiscal Years 2006 to 2012

Dollars in thousands									
Region number[a]	**Region name**	**2006**	**2007**	**2008**	**2009**	**2010**	**2011**	**2012**	**Total**
1	Northern	$8,547	$8,245	$9,804	$9,699	$9,682	$10,799	$9,699	**$66,475**
2	Rocky Mountain	4,352	5,878	6,296	6,525	8,513	8,599	7,308	**47,471**
3	Southwestern	2,637	2,604	3,004	3,238	3,232	3,422	3,108	**21,245**
4	Intermountain	4,862	4,830	5,185	6,334	7,522	7,612	6,634	**42,979**
5	Pacific Southwest	5,683	6,224	6,846	6,701	6,789	7,651	6,701	**46,595**
6	Pacific Northwest	6,052	5,700	6,198	6,234	6,223	7,422	6,296	**44,125**
8	Southern	6,210	6,337	7,330	7,511	7,148	7,395	6,922	**48,853**
9	Eastern	4,913	4,782	4,937	5,566	5,906	5,890	5,478	**37,472**
10	Alaska	3,461	3,208	3,593	3,626	3,812	3,831	3,562	**25,093**
Total		**$46,717**	**$47,808**	**$53,193**	**$55,434**	**$58,827**	**$62,621**	**$55,709**	**$380,309**

Source: Forest Service.

Note: Numbers may not total because of rounding.

[a]The Forest Service does not have a region 7.

Table 7. American Recovery and Reinvestment Act Funding in Support of Forest Service Trail Maintenance and Decommissioning Projects, by Region and State

Region numbera	Region name	State	Number of projects	Recovery Act funding	Total
1	Northern	Montana	2	$6,095,000	$6,095,000
2	Rocky Mountain	Colorado	2	2,950,000	2,950,000
3	Southwestern	Arizona	5	1,636,000	
		New Mexico	15	3,548,000	5,184,000
4	Intermountain	Nevada	3	540,000	540,000
5	Pacific Southwest	California	6	19,123,000	19,123,000
6	Pacific Northwest	Oregon	6	13,285,000	
		Washington	4	2,332,000	15,617,000
8	Southern	Alabama	1	1,415,000	
		Florida	2	1,200,000	
		Georgia	1	751,000	
		Kentucky	2	758,200	
		Mississippi	1	600,000	
		North Carolina	4	6,200,000	
		Puerto Rico	2	900,000	
		Tennessee	2	850,000	
		Virginia	3	3,265,000	15,939,200
9	Eastern	Eastern Regionb	2	4,046,000	
		Illinois	2	600,000	
		Michigan	3	2,180,000	
		Minnesota	2	2,721,000	
		Missouri	1	200,000	
		New Hampshire	1	100,000	
		New York	1	850,000	
		Pennsylvania	1	2,900,000	13,597,000

Region numbera	Region name	State	Number of projects	Recovery Act funding	Total
10	Alaska	Alaska	9	13,382,000	13,382,000
1 or 4c	Idaho	Idaho	7	9,205,000	9,205,000
Total, all regions			90	$101,632,200	$101,632,200

Source: GAO analysis of Forest Service data.

[a] The Forest Service does not have a region 7.

[b] These Eastern Region projects served more than one state. For example, one project supported high-priority trail projects in Illinois, Minnesota, Ohio, and Vermont.

[c] American Recovery and Reinvestment Act funds were distributed by state. Idaho includes portions of both the Northern and Intermountain regions, but the Forest Service budget official providing this information did not know which region the Recovery Act funding supported.

End Notes

[1] GAO, *Parks and Recreation: Maintenance and Reconstruction Backlog on National Forest Trails,* GAO/RCED-89-182 (Washington, D.C.: Sept. 22, 1989). We recommended that the Forest Service gather and make available to Congress, on a periodic basis, nationwide data on needed trail maintenance and reconstruction work, the severity of conditions requiring the work, and associated costs. Since we issued that report, the Forest Service has implemented a database to collect and maintain data related to its trail inventory and to provide data for the agency to estimate the cost of maintenance and reconstruction needs.

[2] A forest trail is one that the Forest Service determines is necessary for the protection, administration, and use of the national forests. The Forest Service's trail mileage has fluctuated over time (see app. II), in part because, according to a Forest Service official, data collection on existing trails has been improved by the agency and, in part, because new trails have been added and existing trails have been decommissioned.

[3] Pub. L. No. 94-588, 90 Stat. 2949 (1976), as amended.

[4] Pub. L. No. 88-577, 78 Stat. 890 (1964), as amended, codified at 16 U.S.C. §§ 1131- 1136.

[5] Pub. L. No. 90-543, 82 Stat. 919 (1968), as amended. The federal portion of the trails system consists of 30 national trails (11 scenic and 19 historic trails, both of which categories must be designated by Congress) covering more than 60,000 miles, with over 1,000 recreation trails. The Forest Service administers 6 national scenic and historic trails: Arizona National Scenic Trail, Continental Divide National Scenic Trail, Florida National Scenic Trail, Nez Perce (Nee Mee Poo) National Historic Trail, Pacific Crest National Scenic Trail, and Pacific Northwest National Scenic Trail. The Forest Service also manages 15 other national scenic and historic trails in cooperation with the Department of the Interior, state and county governments, and private entities.

[6] 16 U.S.C. § 1242(a)(2).

[7] Pub. L. No. 92-300, 86 Stat. 147 (1972), as amended.

[8] According to an agency official, the Forest Service does not track how many volunteer and challenge cost-share agreements are signed at the local level each year. The agency does track national cost-share agreements and reports on them in the agency's annual budget justification.

[9] Under the Volunteers in the National Forests Act of 1972, volunteers are considered federal employees for tort claims purposes.

[10] Under challenge cost-share agreements, the partner's share can be in the form of cash, volunteer services, or in-kind contributions such as equipment.

[11] The Forest Service employs a number of different youth crews, such as those from the Student Conservation Association and Northwest Youth Corps, to maintain trails. In addition to groups from outside the agency, the Forest Service also supports its Youth Conservation Corps program, an outdoor summer job and training program designed for youth aged 15-18.

[12] Act of June 30, 1914, 38 Stat. 430, as amended by Pub. L. No. 104-127, Title III, § 372, 110 Stat. 1015 (1996).

[13] These estimates are based on a sample of surveyed trails and are therefore subject to sampling error, which is usually expressed as a margin of error or confidence interval. Although the Forest Service did not provide the margin of error for its 2012 estimate in its most recent budget justification to Congress, the agency calculated an 80 percent confidence interval that resulted in a margin of error of 4.3 percent (±$23 million) of the estimate. A Forest

Service official confirmed with us that a 95 percent confidence interval, which is a more typical confidence level, would be 6.6 percent (±$34 million) of the estimate.

[14] According to agency officials, surveys are to be conducted annually on 1 percent of trail miles for trail classes 1-4 and on 20 percent of trail miles for class 5 (312 miles of over 158,000).

[15] Land-measuring wheels—consisting of a wheel mounted on a light frame—are used to determine distances, area, and acreages. As a surveyor pushes the frame, a meter records each revolution of the wheel.

[16] GAO, *Federal Real Property: Government's Fiscal Exposure from Repair and Maintenance Backlogs Is Unclear*, GAO-09-10 (Washington, D.C.: Oct. 16, 2008). We did not quantify the extent to which delaying maintenance on trails might increase costs.

[17] According to agency documents and an agency official, the Forest Service receives appropriations by program and does not receive a general administration or management appropriation. As a result, the agency charges each program its share of indirect and support costs through "cost pools." These indirect and support funds pay for district rangers' salaries, administrative support, information technology equipment and support, human resources, rent and utilities, and other common services. The amount each Forest Service program contributes toward these expenses is based on the number of each program's full-time-equivalent staff.

[18] According to officials, the Forest Service intends to complete the travel management planning process by the close of fiscal year 2014. At the time of our review, about 84 percent of forests had completed the planning process, according to these officials.

[19] Some of the variation in the amounts of trails allocations distributed to the regions can be attributed to some regions' receiving funds to help administer and manage national scenic and historic trails. In fiscal year 2012, the agency allocated $7.5 million of its $82 million trails allocation to support national scenic and historic trails.

[20] Information about the number of visitors to each national forest comes from the National Visitor Use Monitoring program, which, according to agency documents, the Forest Service implemented to estimate the volume and characteristics of recreation visitation to the National Forest System, as well as the benefits recreation brings to the American public.

[21] Forest Service officials told us the agency's financial systems do not track how much funding from other programs supports trail maintenance.

[22] Burned area emergency response activities are postfire actions on National Forest System lands to prevent or minimize unacceptable erosion and loss of soil productivity, deterioration of water quality and downstream damage, changes to ecosystem function, establishment of nonnative invasive species, and degradation of cultural and natural resources.

[23] Pub. L. No. 111-5 (2009). These funds were to be obligated by September 30, 2010, and Forest Service headquarters officials told us that nearly 100 percent of the Forest Service's total Recovery Act funds had in fact been obligated by the deadline. The Department of Agriculture's Office of Inspector General has reported on agency trail maintenance-related expenditures under the Recovery Act, including questionable expenditures such as those related to unallowable costs charged by a cooperator. Department of Agriculture, Office of Inspector General, *American Recovery and Reinvestment Act: Forest Service Capital Improvement and Maintenance Projects: Trail Maintenance and Decommissioning*, 08703-0004-SF (Washington, D.C.: July 3, 2012).

[24] The Forest Service began collecting data on volunteer hours for trail maintenance activities beginning in fiscal year 2011; previously, trail maintenance volunteer hours were included as part of volunteer hours counted for all recreation, heritage, and wilderness programs. The agency calculates the value of volunteer labor by multiplying reported volunteer hours by

an hourly rate. Beginning in 2010, the Forest Service based its calculations on a rate for the value of volunteer time estimated by the nonprofit philanthropic network Independent Sector, which was $21.79 per hour in 2011. The Independent Sector describes itself as a "leadership network for nonprofits, foundations, and corporate giving programs." (http://www.independentsector.org/volunteer_time, accessed March 15, 2013).

[25] At the national level, the Forest Service reported receiving nearly $5.9 million in fiscal year 2012 in contributions for trails from partners under challenge cost-share agreements. The agency did not provide information on grants specifically related to trails.

[26] Grants may also be made to private organizations; municipal, county, or state governments; and other federal agencies. The Federal Highway Administration does not track recipients of funding at the state level and is therefore unable to calculate how much Recreational Trails Program funding has supported the Forest Service.

[27] Pub. L. No. 106-393 (2000), as amended. The Secure Rural Schools and Community Self-Determination Act of 2000 was enacted to help address fiscal difficulties confronting rural counties after steep declines in federal timber sales during the 1990s, which significantly decreased revenues to counties from timber sales on national forests managed by the Forest Service and on some public lands managed by the Department of the Interior's Bureau of Land Management. The act, which covers all national forest lands and certain Bureau of Land Management lands in western Oregon, was initially enacted in 2000 and has been reauthorized several times, most recently for a 1-year extension in 2012. As reauthorized, the act comprises three principal titles, the second of which has benefitted Forest Service trail maintenance. In this report, we refer to the Secure Rural Schools and Community Self-Determination Act of 2000 as the Secure Rural Schools Act.

[28] Projects are identified by local resource advisory committees established under Title II of the act. These committees are to contain 15 members representing diverse local interests. For more information on these committees and Title II in general, see GAO, *Update on the Status of Merchantable Timber Contracting Pilot Program*, GAO-10-379R (Washington, D.C.: Mar. 4, 2010).

[29] The National Forest Foundation was created by Congress in 1991 to encourage, accept, and administer private gifts of money and of real and personal property to benefit Forest Service activities.

[30] The agency is unable to determine how many miles of road have been converted to trails as part of the travel management process. It therefore cannot calculate the extent to which road conversions have contributed to the trail maintenance backlog.

[31] The bridge was in good condition when we visited, but officials said that maintaining or replacing the bridge in the future would be expensive and difficult, given the ranger district's current trails allocation of about $300,000-$350,000 each year. Complex bridges can cost from $100,000 to over $1 million to replace, according to officials we interviewed.

[32] Agency officials told us that in some cases, trails can be rerouted without going through a detailed environment analysis.

[33] Section 4 of the Wilderness Act prohibits the construction of temporary roads or structures, as well as the use of motor vehicles, motorized equipment, and other forms of mechanical transport in wilderness areas, unless such construction or use is necessary to meet the minimum requirements for administration of the area, including for emergencies involving health and safety. Generally, the land management agencies have regulations that address the emergency and administrative use of motorized equipment and installations in the wilderness areas they manage.

[34] A crosscut saw is a saw with a handle at each end, designed for use by two people to cut trees across the grain.

[35] The Forest Service uses a process called the minimum requirements decision guide to identify, analyze, and select management actions that are the minimum necessary for wilderness administration. According to a headquarters official, the vast majority of exemptions authorized by the Forest Service are for emergencies, such as wildland fire response or search and rescue, but some exemptions have also been granted for trail maintenance. According to agency data, from 9 to 16 waivers were approved in each of the last 5 years to allow the use of mechanized tools for trail maintenance purposes in designated wilderness. About one-third of these exemptions were for chain saw use; one-third for rock drills; and the remaining third for use of other mechanized equipment, such as generators, wheelbarrows, and battery-powered tools. The agency does not maintain data on requested exemptions, only on those that are approved.

[36] This issue has been much debated, and the Forest Service's Missoula Technology and Development Center in Missoula, Montana, is reviewing current knowledge and issues regarding the safety and efficiency of crosscut saws and chain saws with a report expected by January 2015.

[37] The hemlock woolly adelgid is a small, aphidlike insect native to China and Japan, which is responsible for extensive mortality and decline among hemlock trees in the eastern United States. Laminated root rot is caused by a fungus that decays roots of susceptible conifer trees. As a result, the trees may die because they can no longer take up water and nutrients, or they may be uprooted during windstorms.

[38] Permanent-seasonal employees are guaranteed employment for a certain number of hours each year, according to an agency official. For example, permanent-seasonal employees may be guaranteed full-time employment by the Forest Service for a certain number of 2-week pay periods, followed by a certain number of pay periods when they are on nonpayment status. Temporary positions are for part-time or seasonal employees.

[39] GAO/RCED-89-182.

[40] A Forest Service official told us that public pressure to keep sites open and a lack of ranger district funding to decommission sites led to the closure of few sites. Some stakeholders and officials noted drawbacks to the concept of "right-sizing" the trail system, with stakeholders, in particular, expressing concern that such a process would close trails and decrease access to forests. Moreover, agency headquarters officials also noted that such a process would be expensive and time-consuming because of the required analysis and process for public involvement.

In: Forest Service Trails
Editor: Adam D. Fayden

ISBN: 978-1-62808-834-2

Chapter 2

American Recovery and Reinvestment Act – Forest Service Capital Improvement and Maintenance Projects – Trail Maintenance and Decommissioning*

United States Department of Agriculture

DATE: July 3, 2012
AUDIT NUMBER: 08703-0004-SF
TO: Thomas L. Tidwell, Chief, Forest Service
ATTN: Thelma Strong
Acting Chief Financial Officer
FROM: Gil H. Harden
Assistant Inspector General for Audit
SUBJECT: American Recovery and Reinvestment Act – Forest Service Capital Improvement and Maintenance Projects – Trail Maintenance and Decommissioning

This report presents the results of the subject review. Your written response to the official draft report, excluding the attachments, is included at

* This is an edited, reformatted and augmented version of United States Department of Agriculture Office of Inspector General, dated July 3, 2012.

the end of the report. Excerpts from the response and the Office of Inspector General's (OIG) position are incorporated into the relevant sections of the report. Based on the information in your written response, we have accepted your management decision on Recommendations 1, 2, 4, 5, 6, 7, 8, and 9. Please follow your internal agency procedures in forwarding final action correspondence to the Office of the Chief Financial Officer.

Based on your written response, management decision has not been reached on Recommendation 3. The information needed to reach management decision on this recommendation is set forth in the OIG Position section after the recommendation.

In accordance with Departmental Regulation 1720-1, please furnish a reply within 60 days describing the corrective actions taken or planned, and timeframe for implementing the recommendation for which management decision has not been reached. Please note that the regulation requires management decision to be reached on all recommendations within 6 months from report issuance, and final action to be taken within 1 year of each management decision to prevent being listed in the Department's annual Performance and Accountability Report.

We appreciate the courtesies and cooperation extended to us by members of your staff during our audit fieldwork and subsequent discussions.

Executive Summary

To create jobs and promote economic growth, Congress passed the American Recovery and Reinvestment Act (Recovery Act) on February 17, 2009.[1] The Recovery Act provided Forest Service (FS) $1.15 billion—including $650 million for its Capital Improvement and Maintenance (CIM) fund, and $500 million for its Wildland Fire Management (WFM) fund—to fund projects that directly accomplish FS' mission of sustaining the nation's forests and grasslands. FS allocated $99 million in Recovery Act funds to 90 trail maintenance and decommissioning projects (trail projects). To administer these projects, FS awarded the funds primarily through contracts and partnership agreements.[2] For partnership agreements, FS tended to choose local cooperators (public, private, and non-profit organizations), many of which had an existing partnership with FS.

Congress, in enacting the Recovery Act, emphasized the need for accountability and transparency in the expenditure of funds. Furthermore, the Office of Management and Budget (OMB) issued initial and additional guidance that required Federal agencies to establish rigorous internal controls, oversight mechanisms, and other approaches to meet the accountability objectives of the Recovery Act.[3] The Recovery Act mandated that the U.S. Department of Agriculture's Office of Inspector General (OIG) oversee agency activities to ensure Recovery Act funds were spent in a manner that minimized the risk of improper use. We, therefore, conducted an audit to evaluate FS' administration of the Trail Maintenance and Decommissioning program funds.

We reviewed 24 of the 90 trail projects to determine if FS complied with laws and regulations, selected projects that met eligibility and program requirements, completed the projects in a timely and effective manner, and supported the accomplishments it reported.

For the 24 trail projects, we did not identify any reportable issues related to selection and completion of the projects. We also did not perform a comprehensive review of FS' reported accomplishments because we will address FS' Recovery Act performance measures in a subsequent FS Recovery Act audit.

However, we determined that FS needs to take corrective action to address issues related to compliance with laws and regulations, as we previously reported to agency officials.[4] We found that:

- FS awarded a $9 million youth employment grant with funds from three FS programs without specifying to the grantee the conditions associated with the use of each program's funds.[5] FS personnel stated they overlooked the three different funding sources under the grant because it was quickly awarded.[6] As a result, by December 2009, the grantee had charged FS for various projects unrelated to the three FS programs–we found that 8 of 12 subgrants totaling $317,741 included activities unrelated to the FS programs.[7] FS also arbitrarily allocated over $1.65 million of grant expenditures to the three programs because the grantee was not required to track the expenditures to the three FS programs separately.
- FS program managers at two national forests did not adequately review payment requests from cooperators to ensure that project expenses claimed were for actual and allowable costs.[8] Rather than taking steps to perform additional oversight of the Recovery Act funds, they simply relied on the cooperators to submit accurate claims. As a result, FS overpaid $64,096 in labor costs to one cooperator, and reimbursed another cooperator $24,697 in questionable costs.

Recommendation Summary

To ensure Recovery Act funds are used for their intended purpose and are properly tracked, we recommend that FS provide its grantee with specific direction on the use and tracking of program funds. FS also needs to recover all unallowable costs, and work with the grantee to allocate actual grant expenditures to the three FS programs. We further recommend that FS require its program managers to review supporting documentation to ensure payments to its cooperators are for actual and allowable costs.

Agency Response

In its response dated April 2, 2012, FS generally concurred with the reported findings and recommendations. FS' response, excluding the attachments, is included at the end of this report.

OIG Position

We accept FS' management decision on Recommendations 1, 2, 4, 5, 6, 7, 8, and 9. The actions needed to reach management decision on Recommendation 3 are provided in the OIG Position section after the recommendation.

Background and Objectives

Background

Congress passed the American Recovery and Reinvestment Act (Recovery Act) to create jobs and promote economic growth.[9] Through the Recovery Act, Congress provided Forest Service (FS) with $1.15 billion to fund projects in two major categories: $650 million for Capital Improvement and Maintenance (CIM) and $500 million for Wildland Fire Management (WFM). In response to the Recovery Act's emphasis on using funds as quickly as possible consistent with prudent management, FS approved the first 10 percent of its

Recovery Act funds (about $100 million) in March 2009 for projects that could be quickly started.

FS allocated $99 million of the CIM funds to 90 trail maintenance and decommissioning projects.[10] Of the remaining $551 million, $264 million was used for road, bridge, and related watershed restoration, $247 million for facility maintenance, $25 million for the remediation of abandoned mines, and $15 million for administrative expenses. Chart 1 shows the percentage of funds allocated to each project type within the CIM category.

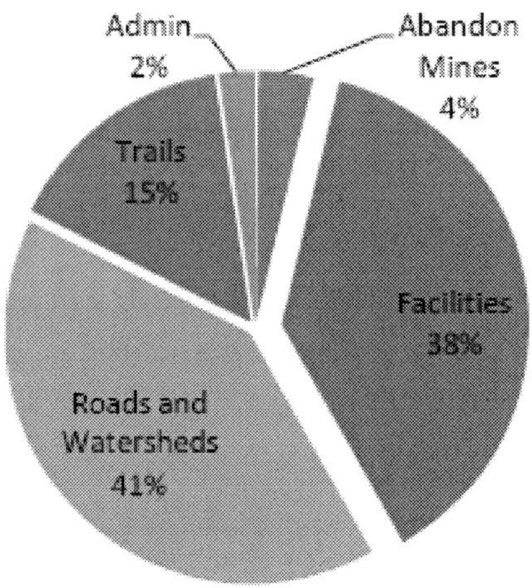

Chart 1. Funding by CIM Project Type.

FS primarily disbursed trail funds through contracts and partnership agreements (including participating agreements[11] and cost-share agreements[12]), with the exception of one grant. For partnership agreements, FS tended to choose local cooperators (public, private, and non-profit organizations), many of which had an existing partnership with FS.

To facilitate fund distribution and provide additional oversight for the Recovery Act projects, FS established four Economic Recovery Operations Centers in April 2009. FS staffed these centers with FS contracting officers and grant and agreement (G&A) specialists. The contracting officers and G&A specialists were responsible for executing and administering the contracts, grants, and agreements for the approved Recovery Act projects. Specifically, the G&A specialists reviewed the grant and agreement applications, determined the proper award type and provisions, and ensured the appropriate authority was cited for the project.

The Recovery Act requires the U.S. Department of Agriculture's (USDA) Office of Inspector General to oversee FS' (and other USDA agencies') activities in order to ensure Recovery Act funds are spent in a manner that minimizes the risk of improper use. Consequently, OIG initiated several audits related to FS' oversight and controls over different programs including the Trail Maintenance and Decommissioning program.

Objectives

Our audit was to determine whether FS (1) complied with laws and regulations pertaining to the Recovery Act funding; (2) selected Recovery Act trail projects that met eligibility and program requirements; (3) timely and effectively completed its Recovery Act trail projects, and (4) can support the accomplishments it reported. For the fourth objective, we did not complete a comprehensive review of FS' reported accomplishments because Recovery Act performance measures will be further addressed in a subsequent FS Recovery Act audit.

Finding 1: FS Did Not Structure a Grant Correctly

FS awarded a $9 million youth employment grant with funds from three FS programs without specifying to the grantee the conditions associated with the use of each program's funds. Because the project was among the first 10 percent of approved Recovery Act projects that needed to be awarded quickly, FS personnel responsible for the grant award overlooked the three different funding sources under the grant.[13] As a result, the grantee charged FS for various activities unrelated to the three FS programs – 8 of 12 subgrants totaling $317,741 included activities unrelated to the FS programs. Further, because the grantee was not required to track the expenditures to the three FS programs separately, FS allocated nearly $1.65 million of grant expenditures to the three programs arbitrarily.

OMB guidance for grant and agreement awards requires agencies to "include terms and conditions in award documents necessary for effective implementation of Recovery Act data collection and accountability requirements."[14] It also states that timely and accurate reporting by Federal agencies provides both Congress and taxpayers an ability to track and monitor all Recovery Act funds with the level of transparency and accountability envisioned in the Act.[15]

To put young adults to work across the State of Oregon, Mt. Hood National Forest awarded a $9 million grant to the Department of Community College and Workforce Development (the "grantee") under the Oregon Youth Employment Initiative project. The project was approved under the first 10 percent of projects selected for funding. FS funded the grant from three different FS programs:

- $5.9 million from the Trail Maintenance and Decommissioning program, a Capital Improvement and Maintenance (CIM) program;
- $2.6 million from the Hazardous Fuels Reduction on Federal Land program, a Wildland Fire Management (WFM) program; and
- $650,000 from the Hazardous Fuels Reduction on non-Federal Land program, a WFM program.

At the time of our review in December 2009, FS had reimbursed the grantee over $1.65 million for youth employment projects in 32 Oregon counties.

FS did not structure the grant according to the above funding allocation and associated program requirements. Instead, the grant stated that the funds were for youth employment projects that fell into the category of natural resources conservation and restoration. The grantee thus allowed its subgrantees to perform a wide range of activities, many of which were not in the scope of the three FS programs. Furthermore, the grantee treated the grant as a single program fund from FS, so the grantee did not track the grant spending for each FS program separately; consequently, FS arbitrarily allocated the grant expenditures to three different programs.

Grant Funds Were Used for Unrelated Activities

FS did not correctly structure the grant to ensure funds were used for intended purposes. Recovery Act Trail Maintenance and Decommissioning program funds were to be invested in "public safety and backcountry access through the operation, maintenance, rehabilitation, and improvement of the National Forest System (NFS) trails system." Recovery Act Hazardous Fuels Reduction funds were to restore fire-adapted ecosystems and to reduce the volume of hazardous fuels on forests, woodlands, shrublands, and grasslands.[16] However, our review found that many projects performed by the subgrantees included activities such as local community services, facility renovations, and trail maintenance on non-NFS land. These activities were not within the scope of the three FS programs.

Specifically, we judgmentally selected and reviewed 12 of the 58 subgrants that FS reimbursed the grantee from April to December 2009 for about $560,000. The accomplishment reports for 8 of 12 subgrants (totaling $317,741) contained activities unrelated to the three FS programs' missions.[17] The unrelated activities performed by the subgrantees included (see exhibit B for more details):

- Community services activities, such as landscaping for homeless drop-in centers and public schools; grounds-keeping for a senior assisted living center; and packing and moving classroom supplies from one elementary school to another.
- Facility renovations such as painting historical buildings for a county fairground, painting a new office building for the community, and repairing buildings at several county fairgrounds.
- Non-NFS trail projects, such as maintaining and improving trails at city, county, and State parks.
- Purchase of several Apple iPhones ®.

The grantee director stated that he was unaware that the grant was funded by three different FS programs. He believed the grant used a single fund for the purpose of promoting conservation education while focusing on youth employment.

We discussed the above activities with FS Pacific Northwest region officials. Although they agreed to provide the grantee with specific program direction regarding CIM trail and WFM hazardous fuels reduction programs, they disagreed that recovery of the reimbursements unrelated to the three FS programs was necessary or appropriate. They claimed a majority of the above unrelated activities enhanced the ecosystem and benefited the watersheds. Specifically, they stated that those activities "improve water quality, quantity, and stream flows in a priority river basin associated with NFS lands on which trail maintenance projects are taking place." They cited FS Recovery Act guidance for other programs (the WFM Ecosystem Improvement Non-Federal Lands program, and the CIM Related Watershed Restoration/Ecosystem Enhancements program) to justify the unrelated activities.

We disagree with regional officials on this point; FS did not fund the grant under these other programs as noted by the specific program codes assigned for the Oregon Youth Employment Initiative project prior to the award process.

FS funded each program for a specific purpose. The use of trail funds for unrelated work did not help the region to address its $46 million backlog of deferred trails maintenance.[18] Similarly, the use of hazardous fuels reduction funds for unrelated activities also did not meet the program purpose of reducing the effects of large destructive fires.

FS Arbitrarily Allocated Grant Expenditures to Three Programs

FS did not correctly charge the grant expenditures to each of the three FS programs. This occurred because the FS program manager considered the grant to be a single fund to create youth employment. As a result, we found FS arbitrarily allocated over $1.65 million in grant expenditures to the three FS programs.

The FS Recovery Act guidance requires Recovery Act funds to be allocated and tracked accordingly to each program.[19]

Based on the allocation of funds from the three programs, 65 percent of the grant's funds (nearly $6 million) were from Trail Maintenance and Decommissioning, 28 percent of the grant funds ($2.6 million) were from Hazardous Fuels Reduction on Federal Land, and 7 percent of the grant funds ($650,000) were from Hazardous Fuels Reduction on non- Federal Land. However, since FS did not inform the grantee of the funding structure and associated program requirements, the grantee did not disburse the funds to the subgrantees according to each program and subgrantees combined their expenditures for activities into a lump sum.

Consequently, the FS program manager assumed that he could record the grant expenditures against any of the three program obligations in the financial system. We observed that by December 2009, FS had arbitrarily allocated over $1.65 million in grant expenditures to the three FS programs.

When we reviewed nine payment requests that FS reimbursed by December 2009, we observed that each request was for funds spent on various program and non-program activities. FS assigned each payment to a single program fund in its financial information system rather than allocating it to the three program funds based on actual activities. For example, a $193,841 payment included expenditures for surveying and removing eight acres of noxious weeds on Federal land (activities covered by the Hazardous Fuels Reduction on Federal Land program), but was charged entirely to the Trail Maintenance and Decommissioning program. In another example, a $159,842 payment that included expenditures for weeding and campsite maintenance on non-Federal land (activities covered by the Hazardous Fuels Reduction on non-Federal Land program) was charged entirely to the Hazardous Fuels Reduction on Federal Land program.

The FS program manager responsible for allocating the funds could not provide us a reasonable explanation for the allocation. Since the grantee was not informed that it should track the expenditures to the three FS programs separately, FS could not accurately allocate the grant expenditures to the three FS programs.

We reported the above issue to FS' Chief in a Fast Report dated July 7, 2010.[20] In FS' July 22, 2010, response to our Fast Report, FS agreed to provide the grantee with specific conditions on the use of the three FS program funds and to properly allocate the expenditures. FS also agreed to recover purchases made for specific community services and facility renovation, but FS did not agree to recover purchases under "ecosystem enhancement-related community service" because those purchases were primarily related to safety requirements to work in the forest. We disagreed with the response; the purchases only represented a small portion of our questioned costs. FS needs to recover all labor and associated administrative costs unrelated to the three FS programs.

Recommendation 1

Provide the grantee with specific direction on what projects should be funded by the three FS programs.

Agency Response

FS concurs with this recommendation, and stated in the July 22, 2010, response to the Fast Report, that [on February 11, 2010] corrective action had been implemented by providing the grantee with specific conditions on the use of the three FS program funds.

OIG Position

We accept FS' management decision on this recommendation.

Recommendation 2

Require the grantee to track the grant expenditures according to each program.

Agency Response

FS concurs with this recommendation. The Agency has implemented corrective action by issuance of a letter [dated February 11, 2010] to the grant recipient requiring grant expenditures be tracked according to each program.

OIG Position

We accept FS' management decision on this recommendation.

Recommendation 3

For our questioned costs totaling $317,741, recover all costs unrelated to the three FS programs.

Agency Response

FS generally concurs with this recommendation. The Agency conducted a review of all costs in question and determined $654 of the $317,741 is unallowable costs. The grantee was notified through a letter correspondence of the amount of costs that was determined unallowable. The unallowable costs were offset in future invoices in order for FS to recover the $654 by reducing it from future disbursements.

OIG Position

We do not accept FS' management decision on this recommendation. As we stated in the Fast Report dated July 7, 2010, and in this report, the grant was not funded under either the Related Watershed Restoration/Ecosystem Enhancements program or WFM Ecosystem Improvement Non-Federal Lands program. Using trail and the two hazardous fuels reduction funds for a broad spectrum of watershed restoration and ecosystem improvement did not meet the specific purpose of the three programs funding the grant. Furthermore, we could not verify whether the $654 represented both the material and labor costs for the unrelated activities because the documentation provided was insufficient.

To reach management decision, FS needs to recover all unrelated costs (labor and supplies) and provide detailed support. If the questioned costs are determined to be uncollectible because FS did not specify the conditions for the use of the funds in the grant, consider waiving the uncollectible amount.

Recommendation 4

Work with the grantee to allocate over $1.65 million of grant expenditures to the three FS programs based on the actual activities that occurred in the field.

Agency's Response

FS concurs with this recommendation, and stated in the July 22, 2010, response to the Fast Report, that [on February 16, 2010] corrective action had been implemented by processing an accounting adjustment to reallocate the existing expenditures among the appropriate FS program funds.

OIG Position

We accept FS' management decision on this recommendation.

Recommendation 5

To prevent this issue from reoccurring with other future grants, provide guidance to grant and agreement officials to ensure all awards include clear terms and conditions on how the program funds will be used.

Agency Response

FS generally concurs with this recommendation. All grants issued by the FS include specified terms and conditions/provisions detailing the recipient's responsibilities with regards to how program funds may be used. The sampled grant in question was an ARRA (Recovery Act) specific grant issued to the State of Oregon, Department of Community College and Workforce Development on April 1, 2009, during the first 10% phase of ARRA funds to be awarded.

During this phase, guidance for awarding funding under ARRA was very limited as the push and intent to get funding to cooperators for stimulus purposes was emphasized. Existing guidance at that time included the "Initial Implementation Guidance for the American Recovery and Reinvestment Act of 2009." ARRA specific guidance was being developed simultaneously and not yet available at the time of this award.

Grants awarded during this phase were issued with the standard terms and conditions/provisions included for grant awards, but may have lacked ARRA specified terms and conditions given the guidance did not yet exist. Subsequent ARRA awards included specified provisions, once such provisions were developed and received. This finding does not represent a systemic gap in how funds are awarded; rather it represents an anomaly specific to the 10% of

ARRA funds awarded during the initial ARRA funding announcement. The Federal Financial Assistance Award of Domestic Grants Template was released in December 2011 for G&A specialist's use to ensure uniformity in all ARRA awards.

Per standard FS procedure, all grants issued include specific terms and conditions regarding program fund use. The FS will continue to issue awards in this manner and be in compliance with expectations. Effective 3/15/2012, should unique funding situations arise akin to the ARRA initiative, whereby after-the-fact specified terms and conditions are developed, executed grants will be modified to include applicable terms and conditions.

OIG Position

We accept FS' management decision on this recommendation.

Finding 2: FS Program Managers Did Not Adequately Review the Cooperators' Payment Requests

FS program managers at two national forests did not adequately review payment requests from cooperators to ensure that project expenses claimed were for actual and allowable costs.[21]

Rather than taking steps to perform additional oversight of the Recovery Act funds, they simply relied on the cooperators to submit accurate claims. As a result, we found FS overpaid $64,096 in labor costs to one cooperator. FS also reimbursed another cooperator $24,697 in questionable costs which included $17,000 for improvements to a new office.

To enhance the accountability and transparency of Recovery Act funds, OMB guidance requires the agencies to take steps, beyond standard practice, to initiate additional oversight mechanisms in order to mitigate the unique implementation risks of the Recovery Act.[22] The FS program manager is responsible for administering the agreement, which includes monitoring the project activities, reviewing the cooperator's payment requests, and assessing the need for corrective action to address performance issues or internal control issues related to the cooperator's financial management.[23]

FS program managers did not always ensure that the payment requests they approved complied with the terms of the participating agreements. Specifically, one program manager approved a cooperator's payment requests even though those requests were based on estimated costs rather than actual costs for the project. Another program manager approved a cooperator's payment requests, all of which contained expenditures unrelated to the Recovery Act trail projects. These issues are discussed in detail below.

Overpayment to CCC

The program manager at Los Padres National Forest approved the payment requests from a cooperator, California Conservation Corps (CCC), even though the expenses claimed were based on estimated costs rather than actual costs. This occurred because rather than taking steps to ensure FS' reimbursements complied with the terms of the agreement, both FS and CCC staff responsible for the Recovery Act trail project relied on an incorrect past practice in which FS always paid CCC an agreed-upon estimated rate. As a result, FS overpaid CCC $64,096 for labor costs.[24]

The Recovery Act-funded participating agreement signed by FS and CCC stated "The U.S. Forest Service shall reimburse the Cooperator for the U.S. Forest Service's share of actual expenses incurred."[25] In addition, the FS Handbook states "the Forest Service is permitted to reimburse only those actual costs (direct and indirect) incurred by the cooperator in performing the project."[26]

FS awarded CCC a $642,000 Recovery Act funded trail project in Los Padres National Forest under a participating agreement.[27] FS paid five CCC payment requests totaling $460,000, which were billed based on estimated costs rather than actual costs.

Our concern prompted CCC headquarters to review its charges to FS. CCC determined that it overcharged FS $64,096 for labor costs.[28] CCC's Administrative Service Division Chief, who is responsible for the agreements, explained that the overbilling error occurred because the local CCC project coordinator had mistakenly used estimated rates rather than actual project costs to bill FS.

The FS program manager explained that FS had partnered with CCC for many years and he had inherited the incorrect payment practice in which FS always paid CCC an agreed-upon estimated rate.[29] Therefore, he overlooked the payment terms stated in the Recovery Act- funded participating agreement. In addition, he was unaware of FS' policy that the cooperator should be reimbursed based on actual project costs.

On August 12, 2011, we reported the above issue to FS' Chief in a Fast Report.[30] FS agreed to provide specific notice requiring FS program managers to ensure payments for participating agreements are based on actual costs and not on an estimated rate. FS also concurred with our recommendation and recovered $64,096 from CCC.

Questionable Costs Charged by a Non-Profit Organization

Gila National Forest did not ensure that it only paid the cooperator, a non-profit organization, for the costs incurred for two trail maintenance projects — FS allocated approximately $370,000 for a 185 mile trail maintenance project and $273,000 for a 145-mile trail maintenance project. This occurred because the program manager did not review supporting documents before approving the organization's payment requests. As a result, FS paid the organization $24,697 in questionable costs from September 2009 to April 2010.

According to FS policy, monitoring is an integral part of post-award administration. The purpose of monitoring is to ensure a cooperator meets the programmatic and financial requirements of the award. The FS program manager is required to review and ensure financial compliance with the terms of the agreement and relevant laws and regulations.[31]

At Gila National Forest, we reviewed two Recovery Act-funded participating agreements totaling $643,000.[32] The organization used Recovery Act funds for unallowable costs, including non-project related costs. Specifically, we determined that the organization used trail funds for new office improvements. We reviewed all the expense records totaling $306,121 for the two agreements and found that the organization spent approximately $17,000 on new office improvements. Expenditures included labor and materials such as wire, lumber, dry wall, custom glass, cabinets, and drawers. The office improvements were not specified in the financial plans of the two Recovery Act-funded agreements.[33] The organization was also unable to provide any documents showing that FS had agreed to pay for the organization's office improvements.

OMB states that "capital expenditures for improvements to land, buildings, or equipment which materially increase their value or useful life are unallowable as a direct cost *except* with the prior approval of the awarding agency;" furthermore, "equipment and other capital expenditures are unallowable as indirect costs."[34] FS payment procedures also state that cooperators shall obtain prior approval of expenditures not included in the agreement financial plan and are required to report deviations from budget and program plans in accordance with the applicable Code of Federal Regulations.[35]

Additionally, as part of our review, we examined the organization's accounting records. We discovered several accounting errors involving the two Recovery Act-funded agreements that were not identified and corrected before FS reimbursed the organization. For example:

- Some purchase requests did not identify the project or program associated with the purchase. We also found instances where unrelated purchases were coded as project expenses. Furthermore, when non-Recovery Act-funded purchases were reclassified as Recovery Act-funded expenditures, there was insufficient evidence to support the reclassification.
- The organization did not always include an invoice number to track its expenditures in the accounting system. Because this unique invoice number was omitted, the organization's system could not detect double accounting entries when expenditures were entered twice.
- The organization did not establish an allocation method for expenses shared between Recovery Act and non-Recovery Act-funded projects.
- In addition, the organization's method to allocate salaries for its Recovery Act- funded trail crews was inconsistent and unsupported.

The accounting errors, which included costs for non-Recovery Act projects and repeated charges for the same purchases, resulted in FS paying the organization $7,699 in questionable costs.

The FS program manager explained that he approved the organization's payment requests without reviewing detailed support for the expenditures because the forest had partnered with the organization on prior trail projects. The program manager told us that he would not have approved the payments if he had known that the organization charged FS for costs unrelated to the two projects.

On December 30, 2010, we reported this issue to FS' Chief in a Fast Report.36 In its written response (January 5, 2011), FS concurred with our recommendations and recovered $17,873 from the organization. FS also agreed to review supporting documents before approving the cooperator's payment requests to ensure that the costs charged to FS are allowable.

Recommendation 6

Recover from CCC the $64,096 overpayment identified at Los Padres National Forest.

Agency Response

FS concurs with this recommendation regarding CCC's overpayment, and stated in the August 25, 2011 response to the Fast Report, that the agency has recovered $64,096 from CCC [on May 16, 2011].

OIG Position

We accept FS' management decision for this recommendation.

Recommendation 7

Recover $24,697 questionable costs identified at the non-profit organization.

Agency Response

FS concurs with this recommendation, regarding Gila National Forest's questionable costs, and stated in the January 5, 2011, response to the Fast Report that the Agency recovered $17,873 from the organization [on October 8, 2010].

Regarding the remainder of $6,824, the Agency reviewed the detailed documentation and noted that some of the questionable costs were allowable administrative costs.

OIG Position

We accept FS' management decision for this recommendation.

Recommendation 8

Issue specific notice reminding FS program managers to ensure payment requests for participating agreements are based on actual costs.

Agency Response

FS concurs with this recommendation, and stated in the August 25, 2011, response to the Fast Report, that the Agency will issue additional guidance to its program managers regarding reimbursements over cooperative agreements [by August 25, 2011].

OIG Position

We accept FS' management decision for this recommendation.

Recommendation 9

To ensure both Recovery and non-Recovery funds are used for the intended purpose, issue specific notice requiring FS program managers to sample and review the supporting documentation against the payment request.

Agency Response

The FS generally concurs with this recommendation. The FS issued guidance to Program Managers by letter on June 2, 2011 regarding necessary steps that must be taken when reviewing and approving requests for payments. The letter was also published on the FS website, providing additional detailed guidance specifically identifying roles and responsibilities of the involved parties.

Performance of appropriate review and maintenance of required file documentation was provided to highlight the importance of complying with ARRA requirements.

FS will issue additional notification (via letter) to make Program Managers aware of their responsibilities to review all supporting documentation as required by FSH 1509.11, section 15.2, FSM 1580.41, and the applicable OMB circulars.

Additionally, in response to a Recovery Act audit, effective July 14, 2011, Grants and Agreements Specialists were notified of criteria for assessing the financial viability of new recipients of Federal Financial Assistance. The letter served as a reminder that it is the responsibility of the Grants Management Specialists to review the Financial Assistance application to ensure applicant's financial strength and capabilities are acceptable, as stated in Forest Service Handbook 1509.11, section 23.52.

A checklist was provided to assess and capture grantee's management capabilities as one of the evaluative criteria used in the administrative review process prior to issuance of an award.

OIG Position

We accept FS' management decision for this recommendation.

SCOPE AND METHODOLOGY

To determine if Recovery Act-funded trail projects met the accountability and transparency objectives of the Recovery Act, we conducted a nationwide review of FS Recovery Act trail maintenance and decommissioning projects.

FS allocated $99 million to 90 trail projects under the Recovery Act. We selected 24 trail projects (4 judgmentally selected and 20 randomly selected) for review (see exhibit D).

We conducted a site visit at 4 Economic Recovery Operations Centers, 2 regional offices, and 20 national forests; we also visited 2 cooperators' sites and 1 grantee's site (see exhibit C). Due to weather conditions that made the project sites inaccessible, we performed desk reviews for 6 selected trail projects located at 8 national forests. We performed our audit work from September 2009 through October 2011.

Our review ultimately included a stratified sample having two strata: a judgmental stratum and a random stratum (see exhibit E). We initially drew random numbers for all projects to select a simple random sample of projects. During the preliminary phase of our audit in January 2010, however, we determined that few projects had sufficient expenditures for our testing. Therefore, we chose first to review five projects based on the criterion that a project's expenditures reached at least 20 percent of its total estimated amount. Due to an issue identified in Gila National Forest, we expanded our judgment stratum to include an additional project, for a total of six projects.

In April 2010, we elected to expand the number of projects reviewed and to employ the random sample methodology. Because two of the six projects in the preliminary review were selected in our random sample,[37] our judgmental sample stratum was reduced to four projects. Our final stratum of four judgmentally-selected projects consisted of nine agreements and one grant for over $13 million.[38]

Our final universe for the random stratum consisted of 86 projects, including 320 contracts and 171 agreements. Our selection used a two-stage approach: the project stage, and the contract and agreement stage. For the first stage, we randomly selected 20 projects.

For the second stage, if our selected project consisted of more than five contracts or five agreements, we then randomly selected a minimum of five from each category. For our 20 selected projects, we reviewed 42 of 72 contracts, and 33 of 46 agreements for nearly $15 million.

We chose to report actual results rather than projections because we observed very few exceptions. For the sample size employed, this low error rate resulted in poor precision for the projected values.

To accomplish our audit, we performed the following steps and procedures:

- Reviewed applicable laws, regulations, policies, and procedures pertinent to the Recovery Act including the Recovery Act of 2009, OMB guidance, the Federal Acquisition Regulation, and FS guidance.
- Interviewed FS regional officials to understand the project selection process, fund disbursement process, and accomplishment reporting process.

- Interviewed the Economic Recovery Operations Center and national forest officials to understand the award and post-award process for execution and administration of contracts, grants, and agreements.
- Reviewed and analyzed 42 contracts, 42 agreements, and 1 grant with associated project files at both Economic Recovery Operations Centers and national forests to determine if FS awarded and administered the projects properly.
- Visited the project sites based on accessibility of the sites, interviewed the personnel, and observed the trail work.
- Interviewed select cooperators and the grantee officials to gain a better understanding of their program operation and project performance.
- Visited selected grantee and cooperators' sites to review their financial records, assess the controls over their accounting system, and examine their project progress reports to determine whether the Recovery Act funds were used for authorized purposes.

We conducted the audit in accordance with generally accepted government auditing standards. Those standards require that we plan and perform the audit to obtain sufficient, appropriate evidence to provide a reasonable basis for our findings and conclusions based on our audit objectives. We believe that the evidence obtained provides a reasonable basis for our findings and conclusions based on our audit objectives.

ABBREVIATIONS

CCC	California Conservation Corps
CIM	Capital Improvement and Maintenance
FS	Forest Service
FSH	Forest Service Handbook
G&A	Grants and Agreements
LPNF	Los Padres National Forest
NFS	National Forest System
OIG	Office of Inspector General
OMB	Office of Management and Budget
USDA.	Department of Agriculture
WFM	Wildland Fire Management

EXHIBIT A: SUMMARY OF MONETARY RESULTS

Finding Number	Recommendation Number	Description	Amount	Category
1	3	Grant Funds Used for Unintended Purposes	$317,741	Questioned Costs Recovery Recommended
1	4	Grant Expenditures Incorrectly Allocated	$1,653,228	Other: Accounting Classification Errors
2	6	Overpayment to the Cooperator	$64,096	Questioned Costs Recovery Recommended
2	7	Unallowable Costs Charged by the Cooperator	$24,697	Questioned Costs Recovery Recommended
TOTAL MONETARY RESULTS			**$2,059,762**	

EXHIBIT B: SUBGRANTS USED FOR UNRELATED ACTIVITIES

Subgrant Number	Amount	Unrelated Activities
GRNT0206	$31,141	Worked in community and school gardens including digging posts, building fences, weeding, and planting; painted and landscaped for homeless drop-in center, performed farming projects for local non- profit organic farms, and built a children's play area.
GRNT0309	$47,794	Cleaned children's park and surrounding streets and bike path; worked on watershed restoration projects such as transects set up, dauber mire and photo plot establishment, fish enumeration and capture.
GRNT0311	$25,129	Reconstructed a trail in State park.
GRNT0312	$63,321	Mowed and pruned 4 acres of a school facility, monitored salmon habitat, collected native seed in a school's native plant nursery; purchased several Apple iPhones ®.

Exhibit B. (Continued)

Subgrant Number	Amount	Unrelated Activities
GRNT0329	$32,000	Planted, weeded, harvested vegetable garden for the food pantries; helped set up a fund raiser; cleaned and painted the county fairground before the Fair week started, set up a fair building for motorcycle show; painted, cleaned, and moved furniture for a non- profit organization; conducted research for historical cemeteries.
GRNT0336	$32,000	Painted historic buildings at fairgrounds; cleaned up fairgrounds and did preparation work for a county fair; cleaned and restocked bathrooms for the city, helped pack and load classroom supplies for an elementary school; maintained a trail at a State park.
GRNT0343	$27,019	Weeded a community garden, reseeded city lawns; did improvement projects at an assisted living center, removed debris and spread gravel at a visitor's center, cleaned county fairgrounds after the fair, painted public restrooms; maintained a trail at a county park.
GRNT0344	$59,337	Maintained trails at several county parks.
TOTAL	$317,741	

EXHIBIT C: SITES VISITED

Organization	**Location**
FS Regional Offices Pacific Northwest Region Pacific Southwest Region	 Portland, Oregon Vallejo, California
Economic Recovery Operations Centers Northwest Southwest Intermountain East	 Sandy, Oregon Vallejo, California Golden, Colorado Atlanta, Georgia
National Forests George Washington & Jefferson NF Nantahala NF Pisgah NF Shawnee NF Superior NF Green Mountain NF White Mountain NF Caribou-Targhee NF Humboldt-Toiyabe NF Mt. Hood NF Tongass NF Carson NF Coronado NF Gila NF Los Padres NF Cleveland NF Mendocino NF	 Roanoke, Virginia Asheville, North Carolina Asheville, North Carolina Harrisburg, Illinois Duluth, Minnesota Rutland, Vermont Campton, New Hampshire Idaho Falls, Idaho Sparks, Nevada Sandy, Oregon Sitka, Alaska Taos, New Mexico Tucson, Arizona

Organization	Location
San Bernardino NF Angeles NF Sierra NF	Silver City, New Mexico Goleta, California San Diego, California Willows, California San Bernardino, California Arcadia, California Clovis, California
Cooperator/Grantee Cooperator A Cooperator B State of Oregon, Department of Community College and Workforce Development	Maplewood, Minnesota Silver City, New Mexico Salem, Oregon

EXHIBIT D: TRAIL PROJECTS REVIEWED

No.	Project	Project Name	Estimated Cost
Judgmental Sample			
1	CIM-0415-21T	Caribou-Targhee Forest Trail Decommissioning Protects Resources	$207,543
2	CIM-0521-03	Region-wide Trail Maintenance Project	$4,015,000
3	OTH-0600-1C1	Oregon Youth Employment Initiative	$3,250,000
4	CIM-0306-08T	FY09 Trail Maintenance Agreements	$306,842
Random Sample			
1	CIM-0100-02T	North Idaho Counties Trail Maintenance & Reconstruction	$5,435,360
2	CIM-0231-09T	Veterans Jobs Trail Project	$458,000
3	CIM-0302-23T	Identify, locate and interpret the Camino Real and Old Spanish National Historic Trails	$102,051
4	CIM-0305-05T	Stop 9 Sabino Cyn Trail Bridge	$120,077
5	CIM-0306-14T	FY09 Trail Maintenance Agreements	$411,000
6	CIM-0417-02T	Mt Rose Wilderness Trails Maintenance & Stewardship	$146,000
7	CIM-05-02T	Wilderness Trail Projects	$4,080,000
8	CIM-0603-10T	Gifford Pinchot Trails Deferred Maintenance	$361,666
9	CIM-0808-14T	Repair and Maintenance of the Virginal Creeper Trail Protects Visitors	$1,475,000

Exhibit D. (Continued)

No.	Project	Project Name	Estimated Cost
10	CIM-0811-08T	North Carolina National Forests Trail Bridges Safety Maintenance	$589,000
11	CIM-0811-09T	Roan Mountain Trails Safety Rehabilitation	$316,000
12	CIM-0908-2T	Phase 1 - Shawnee Designated Trail Improvements	$309,000
13	CIM-0909-1	Enhance Recreational Experiences and Minimize Environmental Impacts by Improving Forest Trails in Minnesota	$557,000
14	CIM-0909-11T	Forest-wide Trail Maintenance	$2,308,700
15	CIM-0920-13T	Forest-wide Recreation Trails Deferred Maintenance Reduction	$850,000
16	CIM-0922-1	Enhance Recreational Experiences and Minimize Environmental Impacts by Improving Forest Trails in New Hampshire	$175,298
17	CIM-1005-10T	Petersburg Mountain Trail Group Maintenance for Safety and Accessibility	$1,481,920
18	CIM-1005-12T	Wrangell Boardwalk Trails Deferred Maintenance for Safety and Protection	$1,521,265
19	CIM-1005-15T	Tongass OHV Bridge Replacement & Trail Maintenance for Visitor Safety and Protection	$620,030
20	CIM-1005-3T	Lake Eva Trail Health & Safety Repair	$868,574

Exhibit E: Sampling Methodology for Trails Projects

Objective

This sample was designed to support the audit of whether trails contracts, grants, agreements, and expenditures using Recovery Act funds complied with applicable laws and agency procedures.

Audit Universe

Our universe for this audit was the list of 90 projects with a CRTR [39] program code funded under the Recovery Act as of April 6, 2010. Forest Service representatives verified that the original list of 89 projects was appropriate and complete as of January 7, 2010. An additional project was added before the April 6 cut-off, bringing the total to 90.

Sample Design and Modifications

The final sample design was stratified by projects at the first stage, with contracts and agreements within projects selected at the second stage.

We observed that many projects had, as of January 7, 2010, little or no expenditure activity. That meant that many of the planned audit tests could not be performed yet. Therefore, we decided to create a census stratum of the five projects having the highest percentages expended compared to the projects' estimated amounts. The second stratum comprised the remaining projects.

We expected each project to involve a number of contracts, agreements, and expenditures, which could not be identified until the relevant project records were reviewed at the appropriate Economic Recovery Operations Center. Therefore, we chose to use a two-stage design within both strata, with projects selected at the first stage and contracts or agreements to be selected at the second stage.

We drew a random number for each project in the audit universe.[40] Five of the projects were assigned to Stratum I based on their expenditure ratio. However, two of those five had very low random numbers. When we expanded the audit work to include the random stratum, those low random numbers placed them among the projects selected for review in the second stratum sample. In the second stratum, therefore, we had a total of 20 randomly-selected projects for review.

When another project was added later, we drew a random number for it using the same range in "randbetween" and placed the additional project into the selection order for Stratum II based on that random number. This resulted in the new project becoming one of the 20 in the simple random sample from the second stratum, and the project originally in the 20th position dropped out of the sample.

We had no information on which to base a sample size calculation. In particular, we did not know whether to expect the variance within clusters or the variance between clusters to dominate. Therefore, the total sample of 24 projects is arbitrary.

As the projects were reviewed at the Economic Recovery Operations Center-level, we obtained information on the recipients and the contracts, agreements, and expenditures. That information was used to make second stage selections for review in the audit.

AGENCY'S RESPONSE

Forest Service's Response to Audit Report

Forest Service

Washington Office
1400 Independence Avenue, SW
Washington, DC 20250

File Code: 1430
Date: April 2, 2012
Route To:

Subject: Response to OIG Draft Report No. 08703-04-SF, "The Recovery Act – Forest Service Trail Maintenance and Decommissioning," dated March 2, 2012

To: Gil Harden, Assistant Inspector General for Audit, Office of Inspector General, USDA

Thank you for the opportunity to review and comment on the OIG draft audit report titled, "Forest Service Trail Maintenance and Decommissioning," dated March 2, 2012. The Forest Service generally concurs with the report's findings and recommendations. The agency's comments regarding the status of recommendation numbers 1, 4, 6, 7 and 8, and our plans to implement recommendation numbers 2, 3, 5 and 9 are enclosed.

If you have any questions, please contact Thelma Strong, Acting Chief Financial Officer, at 202-205-1321 or tstrong@fs.fed.us.

/s/ Thomas L. Tidwell
THOMAS L. TIDWELL
Chief

cc: Dianna Capshaw
Erica Y Banegas
George A Sears
Melissa Dyniec
Melissa A Moreira
Lindsey E DeShazer
Elizabeth Donnelly

USDA Forest Service (FS)

Office of Inspector General (OIG)
Draft Audit Report No. 08703-0004-SF
The Recovery Act - FS Trail Maintenance and Decommissioning

Status Update and Closure

OIG Recommendation #1: Provide the grantee with specific direction on what projects should be funded by the three FS programs.

FS Response: FS concurs with this recommendation, and stated in the July 22, 2010 response to the Fast Report, that corrective action had been implemented by providing the grantee with specific conditions on the use of the three FS program funds.

OIG Position: We accept FS' management decision on this recommendation.

Estimated Completion Date: July 22, 2010

FS Actions Completed to date: FS requests closure of this recommendation. The Agency completed corrective action by the issuance of a letter providing the grantee with specific conditions on the use of the Forest Service programs. See **Enclosure A.**

OIG Recommendation 4: Work with the grantee to allocate over $1.65 million of grant expenditures to the three FS programs based on the actual activities that occurred in the field.

FS Response: FS concurs with this recommendation, and stated in the July 22, 2010 response to the Fast Report, that corrective action had been implemented by processing an accounting adjustment to reallocate the existing expenditures among the appropriate FS program funds.

OIG Position: We accept FS' management decision on this recommendation.

Estimated Completion Date: July 22, 2010

FS Actions Completed to date: FS requests closure of this recommendation. The FS completed corrective action processing an accounting adjustment to reallocate the existing expenditures among the Forest Service program funds. See **Enclosure G.**

OIG Recommendation 6: Recover from CCC the $64,096 overpayment identified at Los Padres National Forest.

FS Response: FS concurs with this recommendation regarding CCC's overpayment, and stated in the August 25, 2011 response to the Fast Report, that the agency has recovered $64,096 from CCC.

OIG Position: We accept FS' management decision for this recommendation.

Estimated Completion Date: May 16, 2011

FS Actions Completed to date: FS requests closure of this recommendation. The FS

completed corrective action by requesting reimbursement for the overpayment. The excess funds were recovered on May 16, 2011. See **Enclosure H and I.**

OIG Recommendation 7: Recover $24,697 questionable costs identified at the non-profit organization.

FS Response: FS concurs with this recommendation, regarding Gila National Forest's questionable costs, and stated in the January 5, 2011 response to the Fast Report that the Agency recovered $17,873 from the organization. Regarding the remainder of $6,824, the Agency reviewed the detailed documentation and noted that some of the questionable costs were allowable administrative costs.

OIG Position: We accept FS' management decision for this recommendation.

Estimated Completion Date: October 8, 2010

FS Actions Completed to date: FS requests closure of this recommendation. The FS completed corrective action by the Forest Service identifying $17,872.79 as unallowable and issuing a Bill of Collection. The excess funds were recovered on October 8, 2010. See **Enclosure J.**

OIG Recommendation 8: Issue specific notice reminding FS program managers to ensure payment requests for participating agreements are based on actual costs.

FS Response: FS concurs with this recommendation, and stated in the August 25, 2011 response to the Fast Report, that the Agency will issue additional guidance to its program managers regarding reimbursements over cooperative agreements.

OIG Position: We accept FS' management decision for this recommendation.

Estimated Completion Date: August 25, 2011

FS Actions Completed to date: FS requests closure of this recommendation. The FS

completed corrective action by issuing guidance to Program Managers. See **Enclosure K.**

Management Decision

OIG Recommendation 2: Require the grantee to track the grant expenditures according to each program.

FS Response: FS concurs with and requests closure of this recommendation. The Agency has implemented corrective action by issuance of a letter to the grant recipient requiring grant expenditures be tracked according to each program. See **Enclosure A**.

Estimated Completion Date: February 11, 2010

OIG Recommendation 3: For our questioned costs totaling $317,741, recover all costs unrelated to the three FS programs.

FS Response: FS generally concurs with and requests closure of this recommendation. The Agency conducted a review of all costs in question and determined $654 of the $317,741 is unallowable costs. The grantee was notified through a letter correspondence of the amount of costs that was determined unallowable. The unallowable costs were offset in future invoices in order for FS to recover the $654 by reducing it from future disbursements. See **Enclosures B, C, D, E, and F.**

Estimated Completion Date: August 8, 2010

OIG Recommendation 5: To prevent this issue from reoccurring with other future grants, provide guidance to G&A officials to ensure all awards include clear terms and conditions on how the program funds will be used.

FS Response: FS generally concurs with and requests closure of this recommendation. All grants issued by the FS include specified terms and conditions/provisions detailing the recipient's responsibilities with regards to how program funds may be used. The sampled grant in question was an ARRA specific grant issued to the State of Oregon, Department of Community College and Workforce Development on April 1, 2009, during the first 10% phase of ARRA funds to be awarded. During this phase, guidance for awarding funding under ARRA was very limited as the push and intent to get funding to cooperators for stimulus purposes was emphasized. Existing guidance at that time included the "Initial Implementation Guidance for the American Recovery and Reinvestment Act of 2009." ARRA specific guidance was being developed simultaneously and not yet available at the time of this award.

Grants awarded during this phase were issued with the standard terms and conditions/provisions included for grant awards, but may have lacked ARRA specified terms and conditions given the guidance did not yet exist. Subsequent ARRA awards included specified provisions, once such provisions were developed and received. This finding does not represent a systemic gap in how funds are awarded; rather it represents an anomaly specific to the 10% of ARRA funds awarded during the initial ARRA funding announcement. The *Federal Financial Assistance Award of Domestic Grants* Template was released in December 2011 for G&A specialist's use to ensure uniformity in all ARRA awards. See **Enclosure L.**

Per standard FS procedure, all grants issued include specific terms and conditions regarding program fund use. The FS will continue to issue awards in this manner and be in compliance with expectations. Effective 3/15/2012, should unique funding situations arise akin to the ARRA initiative, whereby after-the-fact specified terms and conditions are developed, executed grants will be modified to include applicable terms and conditions.

Estimated Completion Date: March 15, 2012

OIG Recommendation 9: To ensure both Recovery and non-recovery funds are used for the intended purpose, issue specific notice requiring FS program managers to sample and review the supporting documentation against the payment request.

FS Response: The FS generally concurs with this recommendation. The FS issued guidance to Program Managers by letter on June 2, 2011 regarding necessary steps that must be taken when reviewing and approving requests for payments. The letter was also published on the FS website, providing additional detailed guidance specifically identifying roles and responsibilities of the involved parties. Performance of appropriate review and maintenance of required file documentation was provided to highlight the importance of complying with ARRA requirements.

FS will issue additional notification (via letter) to make Program Managers aware of their responsibilities to review all supporting documentation as required by FSH 1509.11, section 15.2, FSM 1580.41, and the applicable OMB circulars.

Additionally, in response to a Recovery Act audit, effective July 14, 2011, Grants and Agreements Specialists were notified of criteria for assessing the financial viability of new recipients of Federal Financial Assistance. The letter served as a reminder that it is the responsibility of the Grants Management Specialists to review the Financial Assistance application to ensure applicant's financial strength and capabilities are acceptable, as stated in Forest Service Handbook 1509.11, section 23.52. A checklist was provided to assess and capture grantee's management capabilities as one of the evaluative criteria used in the administrative review process prior to issuance of an award.

Estimated Completion Date: December 31, 2012

End Notes

[1] Public Law 111-5 (February 17, 2009).

[2] A voluntary, mutually beneficial arrangement between FS and an organization (public or private) for the purpose of accomplishing mutually agreed upon objective(s).

[3] OMB Memorandum M-09-10 (February 18, 2009); and OMB Memorandum M-09-15 (April 3, 2009).

[4] OIG reported these issues to FS through three Fast Reports: 08703-4-SF(1) "The Recovery Act – FS Trail Maintenance and Decommissioning" (July 7, 2010); 08703-4-SF(2) "Gila National Forest, New Mexico- Questionable Costs-Recovery Act-FS Trail Maintenance and Decommissioning" (December 30, 2010); and 08703-4-SF(3) "Los Padres National Forest, Cost Overpaid to California Conservation Corps – Recovery Act-FS Trail Maintenance and Decommissioning Fast Report" (August 12, 2011).

[5] The grant was funded as follows: about $5.9 million from the Trail program, $2.6 million from the Hazardous Fuels Reduction on Federal Land program, and $650,000 from the Hazardous Fuels Reduction on non-Federal Land program.

[6] The project was one of the first 10 percent of projects selected for funding. These projects were selected because they could be started quickly and were expected to be awarded as soon as possible.

[7] At the time of our review in December 2009, the grantee had awarded 58 subgrants totaling about $2.3 million to 34 Oregon counties.

[8] A cooperator is an individual or entity that voluntarily cooperates with FS on a project and is willing to formalize the relationship by entering into a Memorandum of Understanding or agreement.

[9] Public Law 111-5 (February 17, 2009).

[10] As of September 30, 2010.

[11] FS partners with State and non-profit conservation corps to maintain and enhance national forest trails, as well as to teach young people job skills and conservation principles.

[12] FS cooperates with public and private entities or individuals in developing, planning, and implementing mutually beneficial projects that enhance FS activities.

[13] The grant was awarded under the Oregon Youth Employment Initiative project.

[14] OMB Memorandum M-09-10 Section 1 p.6 (February 18, 2009).

[15] OMB Memorandum M-09-15 Section 1 p.7 (April 3, 2009).

[16] FY 2009 American Recovery and Reinvestment Act (FS guidance), Chapter 2 (March 18, 2009) p. 2-1, 2-2, 2-10, and 2-11.

[17] In our examples, we were unable to determine the specific dollar amount charged to each activity because the grantee did not disburse and track the grant funds according to each program, and subgrantees combined their expenditures for activities into a lump sum.

[18] As of November 25, 2009, the FS Pacific Northwest region's deferred trail maintenance backlog was valued at $46 million.

[19] FY 2009 American Recovery and Reinvestment Act (FS guidance) Chapter 2 p. 2-5 (March 18, 2009), and Chapters 3 and 4 (March 6, 2009) p. 3-2 and 3-3, p. 4-2 to 4-6.

[20] 08703-0004-SF(1) "The Recovery Act – FS Trail Maintenance and Decommissioning" (July 7, 2010).

[21] A cooperator is an individual or entity that voluntarily cooperates with FS on a project and is willing to formalize the relationship by entering into a Memorandum of Understanding or agreement.

[22] OMB Memorandum M-09-10 (February 18, 2009).

[23] Forest Service Manual 1580 p. 23-24 (October 20, 2009).

[24] As of July 2010, CCC claimed $460,000 out of $642,000 awarded by Los Padres National Forest. The overpayment of $64,096 is about 14 percent of the total claims.

[25] Recovery Act Participating Agreement No. 10-PA-11059702-056, p. 3 (November 25, 2009).

[26] Forest Service Handbook (FSH) 1509.11, section 70.3 p. 8 (October 20, 2009).

[27] For a selected Wilderness Trail Project in Southwest Region, we randomly selected five participating agreements and six contracts totaling $1.3 million for review. Three of the five agreements were awarded to CCC by Los Padres National Forest, Mendocino National Forest, and Cleveland National Forest for $642,000, $180,000, and $280,096, respectively.

[28] OIG together with the CCC regional analyst reconciled all of the project expenses with CCC's supporting documentation.

[29] We reviewed the invoices under an ongoing non-Recovery Act agreement between Los Padres National Forest and CCC, and we noted that FS incorrectly reimbursed CCC by using an estimated rate instead of the actual cost.

[30] 08703-4-SF(3) "Los Padres National Forest, Cost Overpaid to California Conservation Corps – Recovery Act- FS Trail Maintenance and Decommissioning Fast Report" (August 12, 2011).

[31] FSH 1509.11, section 10 p. 23-24 (October 20, 2009).

[32] FS allocated approximately $370,000 for the 185-mile project and $273,000 for the 145-mile project.

[33] The financial plan is a spreadsheet attached to an agreement that displays the contribution from each party separated by cost element, which indicates the detailed breakdown of costs anticipated for the project.

[34] OMB Circular No. A-122, Cost Principles for Non-Profit Organizations, Attachment B 15 b (1) and 15 (d) (revised May 10, 2004).

[35] FSH 1509.11, section 74.4 p. 64 (October 20, 2009).

[36] 08703-4-SF(2) "Gila National Forest, New Mexico-Questionable Costs-Recovery Act-FS Trail Maintenance and Decommissioning" (December 30, 2010).

[37] Selections were identified using the originally-drawn random numbers.

[38] The dollar amount included the project at Caribou-Targhee National Forest with no contract or agreement awarded and $2.875 million in CIM trail funds added to the grant through the Oregon Youth Employment Initiative Project (phase II).

[39] CRTR is a program code established by FS to identify all the trail maintenance and decommission projects funded by Recovery Act.

[40] We drew random numbers using the Excel® randbetween (1,999) function.

Index

A

access, vii, 1, 3, 8, 22, 26, 41, 55, 63
accessibility, 76
accountability, viii, 58, 62, 69, 74
accounting, 68, 71, 72, 76, 84
adjustment, 68, 84
administrative support, 53
administrators, 41
agencies, viii, 6, 10, 23, 32, 46, 54, 59, 62, 69
agency decisions, 33
Alaska, 47, 48, 49, 51, 78
American Recovery and Reinvestment Act, v, vii, 18, 21, 48, 50, 51, 53, 57, 58, 60, 68, 86, 88, 89
American Recovery and Reinvestment Act of 2009, 18, 21, 68, 86
appropriations, 10, 18, 53
assessment, 23, 36
audit, viii, 6, 46, 58, 59, 62, 74, 75, 76, 80, 81, 82, 87
audits, 62
authorities, 8
authority, 5, 11, 24, 44, 61

B

barriers, 10
base, 20, 81
beetles, 20, 29
benefits, 45, 53
breakdown, 89
budget line, 18
Bureau of Land Management, 6, 46, 54

C

cabinets, 71
capital expenditure, 71
cash, 52
certification, 33, 39
challenges, 5, 26, 28, 33, 46
children, 77
China, 55
City, 79
classes, 20, 48, 53
classroom, 64, 78
closure, 55, 83, 84, 85, 86
clusters, 81
Code of Federal Regulations, 71
collaboration, 3, 32, 39, 42, 43
collateral, 32, 39, 42
community, 35, 64, 66, 77, 78
community service, 66
compensation, 11, 33, 40
complement, 32, 39
complexity, 29
compliance, 59, 69, 71, 87
complications, 26, 31

conference, 46
Congress, vii, viii, 8, 10, 18, 32, 52, 54, 58, 60, 62
conifer, 55
conservation, vii, 2, 6, 8, 11, 45, 63, 64, 88
construction, 7, 10, 14, 23, 31, 36, 54
Continental, 52
cooperation, 52, 58
cooperative agreements, 73, 85
cost, 11, 16, 17, 20, 22, 23, 25, 27, 28, 29, 36, 40, 52, 53, 54, 61, 89
covering, 11, 52
culture, 29
curriculum, 36
cycles, 10, 47

D

data collection, 14, 30, 37, 41, 52, 62
database, 5, 38, 44, 48, 52
degradation, 53
Department of Agriculture, v, 4, 43, 53, 57, 76
Department of the Interior, 6, 46, 52, 54
Department of Transportation, 44
direct cost, 71
directors, 20
disbursement, 75
diseases, 29
distribution, 38, 61
draft, 2, 57, 82
drainage, 9, 16, 17, 29

E

economic growth, vii, 58, 60
ecosystem, 53, 64, 66, 67
education, 32, 64
elementary school, 64, 78
emergency, 19, 21, 53, 54
emergency response, 21, 53
employees, 8, 22, 30, 32, 33, 35, 37, 41, 43, 52, 55
employment, 55, 59, 62, 63, 64, 65
engineering, 26
environment, 54
epidemic, 20, 29
equipment, 10, 26, 28, 33, 41, 52, 53, 54, 55, 71
erosion, 3, 16, 17, 21, 26, 28, 35, 53
evidence, 5, 6, 44, 46, 72, 76
execution, 76
executive branch, 32
Executive Order, 32
exercise, 38
expenditures, 53, 59, 60, 62, 63, 65, 66, 67, 68, 70, 71, 72, 75, 80, 81, 82, 83, 84, 85, 86, 88
expertise, 3, 26, 30, 37, 42, 43

F

farms, 77
fear, 40
Federal Highway Administration, 23, 44, 54
financial, 25, 53, 65, 69, 71, 74, 76, 87, 89
financial records, 76
financial resources, 25
financial system, 53, 65
fires, 5, 21, 65
fish, 8, 77
Fish and Wildlife Service, 6, 46
flaws, 28
flooding, 29
floods, 29
fluctuations, 47
food, 31, 78
force, 16
foundations, 54
funding, vii, 1, 2, 5, 18, 20, 21, 22, 23, 24, 33, 34, 36, 38, 40, 41, 44, 48, 50, 51, 53, 54, 55, 59, 62, 63, 65, 67, 68, 69, 86, 87, 88
funds, vii, viii, 5, 10, 11, 18, 19, 20, 21, 22, 23, 24, 38, 40, 44, 51, 53, 58, 59, 60, 61, 62, 63, 65, 66, 67, 68, 69, 71, 73, 76, 80, 83, 84, 85, 86, 87, 88, 89
fungus, 55

G

GAO, vii, 1, 2, 3, 13, 17, 19, 21, 27, 45, 46, 51, 52, 53, 54, 55
geography, 5, 44
Georgia, 17, 27, 45, 50, 78
governments, 52, 54
grades, 9
grant programs, 24
grants, 11, 23, 24, 25, 31, 33, 40, 45, 54, 61, 68, 69, 76, 80, 86, 87
grasslands, vii, 4, 6, 11, 58, 63
guidance, viii, 5, 35, 40, 42, 43, 44, 59, 62, 64, 65, 68, 69, 73, 74, 75, 85, 86, 87, 88, 89
guiding principles, 34

H

habitat, 7, 8, 77
harvesting, 3, 8, 26
hazards, 14
health, 54
hiring, 32, 37, 39, 41
horseback riders, vii, 1, 16
horses, 30
House, 4
House of Representatives, 4
human, 9, 53
human resources, 53
hunting, 4
hurricanes, 29

I

improvements, 10, 18, 69, 71
Independence, 82
individuals, 11, 88
industry, vii, 2, 6, 45
information technology, 53
infrastructure, 24, 34, 42
insects, 29
internal controls, viii, 59
investment, 19, 20
investments, 34, 42
issues, 20, 24, 26, 41, 55, 59, 69, 70, 88

J

Japan, 55
job skills, 88
justification, 10, 52

L

laws, vii, 1, 5, 44, 59, 62, 71, 75, 80
laws and regulations, 59, 62, 71
lead, 12, 17
leadership, 7, 30, 54
legislation, 40
liability insurance, 11
light, 5, 34, 42, 53
livestock, 8
local community, 63

M

magnitude, 41
maintenance tasks, 10
majority, 9, 11, 28, 32, 55, 64
management, 3, 6, 7, 8, 9, 10, 11, 20, 23, 26, 32, 34, 36, 37, 39, 42, 43, 45, 46, 53, 54, 55, 58, 60, 66, 67, 68, 69, 73, 74, 83, 84, 85, 88
materials, 25, 71
melts, 28
meter, 53
methodology, 6, 75
Mexico, 45, 50, 78, 79, 88, 89
mission(s), vii, 38, 58, 64
Missouri, 50
models, 38
Montana, 15, 16, 27, 45, 50, 55
mortality, 55

N

National Forest System, 6, 7, 9, 11, 41, 47, 48, 53, 63, 76
national forests, vii, 1, 4, 5, 6, 7, 8, 10, 11, 14, 20, 23, 24, 44, 52, 54, 59, 69, 74, 76
National Park Service, 6, 46
natural resources, 2, 12, 14, 16, 33, 35, 42, 53, 63
negative effects, 2, 14, 16
nutrients, 55

O

obstacles, 9, 16
off-highway-vehicle drivers, vii, 1
Office of Management and Budget, viii, 58, 76
officials, vii, 2, 3, 5, 7, 8, 12, 14, 16, 18, 20, 21, 22, 24, 25, 26, 27, 28, 30, 31, 32, 33, 34, 35, 36, 37, 38, 39, 40, 44, 46, 53, 54, 55, 59, 64, 68, 75, 76, 86
OIG Position, 58, 60, 66, 67, 68, 69, 73, 74
OMB, viii, 58, 62, 69, 71, 74, 75, 76, 87, 88, 89
operations, 2, 12, 14
opportunities, 4, 8, 35, 36, 37, 38, 41
organize, 10
overhead costs, 19, 25
oversight, viii, 59, 61, 62, 69
ownership, 33

P

Pacific, 8, 18, 21, 28, 29, 45, 46, 47, 48, 49, 50, 52, 64, 78, 89
participants, 46
policy, 7, 70, 71
pools, 20, 40, 53
population, 5, 44
preparation, 78
principles, 88
prisoners, 25
profit, viii, 58, 61, 71, 73, 77, 78, 84, 88
progress reports, 76
project, 11, 20, 51, 59, 61, 62, 63, 64, 69, 70, 71, 72, 75, 76, 81, 88, 89
protection, 8, 11, 28, 52
pruning, 12
public safety, 8, 63
public schools, 64
Puerto Rico, 6, 50

Q

quality of life, 4
quality standards, 2, 9, 10

R

rainfall, 18
random numbers, 75, 81, 89
recommendations, 2, 43, 58, 60, 72, 82
reconstruction, 4, 52
recovery, 16, 64, 87
recreation, 3, 4, 8, 20, 26, 34, 52, 53
recreational, vii, 1, 4, 5, 11, 38, 41, 47
recruiting, 37
regulations, 5, 39, 44, 54, 75
rehabilitation, 8, 14, 63
reimburse, 70
reliability, 5, 38, 44, 45
remediation, 61
rent, 53
repair, 9, 14, 22, 32
requirements, 28, 33, 40, 54, 55, 59, 62, 63, 65, 66, 71, 74, 87
reserves, 19
resources, vii, 1, 2, 3, 5, 11, 16, 18, 22, 24, 26, 28, 30, 31, 32, 34, 36, 41, 42, 44
response, 21, 55, 57, 58, 60, 66, 68, 72, 73, 74, 83, 84, 85, 87
restoration, 21, 61, 63, 67, 77
risk(s), viii, 33, 59, 62, 69
root(s), 18, 29, 55
root rot, 29, 55
routes, 9, 16, 23
rural areas, 5, 44
rural counties, 54

S

safety, 14, 22, 26, 31, 33, 41, 54, 55, 66
salmon, 77
sample design, 81
sampling error, 52
scarce resources, 42
school, 77
scope, 6, 63, 64
Secretary of Agriculture, 23, 42
sediment, 16
seed, 77
services, 10, 11, 39, 52, 53, 63, 64
showing, 71
signs, 9, 15, 22, 23
snowmobiles, 4
software, 46
soil type, 17
solitude, 38
specialists, 61
species, 16, 53
spending, 42, 63
staffing, 10
stakeholder groups, vii, 2
stakeholders, 2, 3, 6, 15, 22, 25, 26, 27, 28, 30, 31, 32, 33, 34, 35, 36, 38, 39, 40, 55
state, 9, 23, 24, 45, 48, 51, 52, 54, 71
states, 2, 6, 18, 22, 23, 29, 37, 62, 70, 71
stimulus, 68, 86
storms, 29
structure, 63, 65
subsistence, 10
supervision, 32
supervisor, 7
sustainability, 2, 3, 34, 35, 36, 37, 41, 42

T

target, 12
taxpayers, 62
technical assistance, 10
technical comments, 43
testing, 75
time frame, 35
Title I, 23, 25, 52, 54
Title II, 23, 25, 52, 54
tornadoes, 29
tracks, 23
training, 2, 3, 10, 29, 30, 32, 33, 35, 36, 39, 40, 41, 42, 43, 52
training programs, 30
transparency, viii, 58, 63, 69, 74
transport, 29, 54
transportation, 10
treatment, 39

U

U.S. Department of Agriculture, viii, 59, 62
U.S. economy, 4
U.S. Geological Survey, 6, 46
uniform, 39
United, v, 1, 3, 6, 55, 57
United States, v, 1, 3, 6, 55, 57
universe, 75, 80, 81
up-front costs, 28, 35
urban, 5, 44
USDA, 62, 76, 82, 83

V

vegetation, 9, 18, 22
vehicles, 4, 11, 23, 25, 26, 36, 47, 48, 54
vision, 34, 39, 42
volunteer work, 10
Volunteers, 10, 31, 32, 33, 39, 52

W

walking, 4
Washington, 3, 7, 17, 45, 46, 50, 52, 53, 54, 78, 82
waste, 9
water, 8, 9, 16, 17, 21, 26, 27, 28, 53, 55, 64
water quality, 16, 21, 53, 64
watershed, 8, 21, 61, 67, 77
web, 32, 37, 46
wilderness, 8, 9, 20, 26, 28, 38, 41, 48, 53, 54, 55

wildland, 5, 21, 26, 28, 55
wildlife, 7, 8, 10
windstorms, 29, 55
workers, 11, 33, 37, 40

Y

young adults, 63
young people, 88